SAE J1939 ECU PROGRAMMING

& VEHICLE BUS SIMULATION

WITH ARDUINO

COPPERHILL TECHNOLOGIES CORPORATION

http://www.copperhilltech.com

SAE J1939 ECU Programming & Vehicle Bus Simulation
with Arduino

By Wilfried Voss

Published by
Copperhill Technologies Corporation
158 Log Plain Road
Greenfield, MA 013901

ISBN-10: 1938581180
ISBN-13: 978-1-938581-18-2

http://copperhilltech.com

From the Author

After writing *Controller Area Networking (CAN) with Arduino* (see literature appendix in this book), the next logical step was looking into higher-layer protocols based on CAN, in this case the SAE J1939 vehicle network protocol. I personally consider J1939 the most effective CAN protocol in the market due to its small memory footprint combined with a bandwidth that easily surpasses other protocols such as CANopen and DeviceNet.

However, my swift choice for SAE J1939 was not only based on technical aspects but also on the great popularity the J1939 protocol enjoys in the North American market. CANopen and DeviceNet, in view of the overpowering strength of Industrial Ethernet for automation, are nearing their virtual end-of-lifetime cycle for new developments and for that reason I dismissed any thoughts of writing a book about them.

The Arduino presented itself again as the obvious choice due to its vast popularity, high-level of user-friendliness and, after all, low costs. The programming and implementation of an SAE J1939 protocol stack using the Arduino requires, of course, not only some knowledge of C/C++ programming but also some basic knowledge of the Controller Area Network (CAN) technology and the SAE J1939 protocol. While I am trying to explain details where necessary (i.e. where the information pertains to programming the code), a fully detailed description of either topic is out of the scope of this book.

For further, detailed information on Controller Area Network and SAE J1939 see my books:

- A Comprehensible Guide to Controller Area Network
- A Comprehensible Guide to J1939

Both titles are available in paperback form through Amazon.com (including all their language specific sites), Barnes & Noble (bn.com), Abebooks.com (including all their language specific sites), and other online bookstores all over the world.

In the same sense, I will also not engage into explaining the basics of the Arduino hardware and its programming. There are myriads of books on Arduino, Arduino Sketches, and Arduino Shields available on the topic, and it makes no sense repeating the information therein just to increase the page count.

I will, however, briefly refer to the hardware used in my J1939 projects, namely the Arduino Uno, the Arduino Mega 2560, and the CAN Shield, but only for the purpose of providing information that will enable the reader to replicate my projects. I deem it also important to point to the hardware's performance limitations. After all, the Arduino may be the prefect prototyping solution, but implementing, for instance, a full-blown SAE J1939 protocol definitely meant pushing the limits.

I will also refer to some programming topics that not only go beyond Arduino basics but also reflect my personal approach to writing readable code. These topics include programming style, debugging code, and memory management.

ARD1939 – SAE J1939 Protocol Stack for Arduino

I deem it necessary to add a few non-technical (and maybe politically incorrect) aspects on the development of the most interesting feature of this book, the ARD1939, an SAE J1939 protocol stack for Arduino.

The implementation of an SAE J1939 protocol stack was (and in many cases still is) out of financial reach for many engineers. The software (i.e. the source code) is either grossly overpriced or comes with hefty royalties (object code, libraries), meaning you have to pay for each copy. At least in the case of the Arduino hardware, this is going to change, and the ARD1939 protocol stack project is available as a free download.

Quite honestly, the development of a J1939 protocol is not a big deal for an experienced programmer, the only obstacle being that you have to spend some good money on the official document, the SAE J1939 Standards Collection. I deem my code as good as any commercially available protocol stacks. And yes, I have successfully tested my code against a number of commercially available J1939 devices.

I had contemplated releasing ARD1939 in form of the original source code but ultimately decided against it, mostly out of respect for those small businesses that make a living from selling SAE J1939 devices and software tools. Instead, I provide a pre-compiled code.

The original source code could be easily adapted to any other embedded hardware (I have it already running on an ARM system) and even Windows or Linux PCs with CAN capabilities.

Furthermore, the costs for off-the-shelf, industrial-strength hardware with an extended temperature range are surprisingly low these days. In my experience, you can easily create an SAE J1939 node with USB and Ethernet port (gateway application) with industrial-strength enclosure for under US$200 at volume = 1.

I believe that the average Arduino user, through using the pre-compiled ARD1939 code, will be quite able to write effective J1939 applications, regardless of whether or not he/she has access to the original source code. After all, there is no real need to modify a working code. It supports all SAE J1939 protocol features, and the focus should always be on the actual application.

About the Author

I am the author of the "Comprehensible Guide" series of technical literature covering topics like Controller Area Network (CAN), SAE J1939, Industrial Ethernet, and Servo Motor Sizing. I have worked in the CAN industry since 1997 and before that was a motion control engineer in the paper manufacturing industry. I have a master's degree in electrical engineering from the University of Wuppertal in Germany.

During the past years, I have conducted numerous seminars on industrial fieldbus systems such as CAN, CANopen, SAE J1939, Industrial Ethernet, and more during various *Real Time Embedded And Computing Conferences* (RTECC), ISA (Instrumentation, Systems, and Automation Society) conferences and various other events all over the United States and Canada.

I had the opportunity of traveling the world extensively, but settled in New England in 1989. I presently live in an old farmhouse in Greenfield, Massachusetts with my red-haired, green-eyed Irish-American wife and our son Patrick.

For more information on my works and to contact me, see my website at http://copperhilltech.com.

Contact the Author

Despite all efforts in preparing this book, there is always the possibility that some aspects or facts will not find everybody's approval, which prompts us, the author and the publisher, to ask for your feedback. If you would like to propose any amendments or corrections, please send us your comment. We look forward to any support in supplementing this book, and we welcome all discussions that contribute to making the topic of this book as thorough and objective as possible.

To submit amendments and corrections please log on to the author's website at http://copperhilltech.com/contact-us/ and leave a note.

In addition, register as a user at http://www.j1939forum.com where we have set up an author section. Please use this forum for communication with the author. However, use it only to request information pertaining to the content of this book. Any inquiries requesting help for specific user projects and/or debugging will be ignored.

Download the Code

Each programming example in this book, either Arduino C/C++ Sketches or Visual Studio C# projects, are available as a free download. This also includes the ARD1939 protocol stack project.

To download the sketches, projects and libraries go to:

http://ard1939.com

I contemplated listing all Arduino projects and their code in printing at the end of this book, not as an attempt of increasing the page count, but due to paying respect to those who unable to download the code (yes, they do exist). However, all Arduino sketches use the MPC2515 Library by Cory Fowler (See chapter Arduino Programming (Sketches)), which I cannot provide in printing due to copyright restraints. In all consequence, in order to fully use the code as introduced in this book, you do need an Internet connection for downloads.

All Arduino sketches and other code samples and projects as introduced in this book are free software; you can redistribute and/or modify them. The programs are introduced in the hope that they will be useful, but WITHOUT ANY WARRANTY; without even the implied warranty of MERCHANTABILITY or FITNESS FOR A PARTICULAR PURPOSE. With downloading these programs, you confirm that these code samples and projects were created for demonstration and educational purpose only.

Table of Content

1. Introduction

Any embedded computing project is as much about the hardware as it is about the software. In this particular project, the ultimate goal is to run an SAE J1939 protocol stack (the software) on the Arduino (the hardware).

Only a few years ago, such a project required some major investments in embedded hardware combined with a steep learning curve, but, like in many other areas of embedded programming, the Arduino family of microcontrollers made it tremendously easier to build even complex embedded solutions. Moreover, the availability of various extension boards (Shields), such as a CAN shield, opened the door to endless possibilities.

In the same sense, the use of an SAE J1939 protocol stack was out of financial reach for many engineers. The software (source code) is either grossly overpriced or comes with royalties (object code, libraries), meaning you have to pay for each copy. At least in the case of the Arduino hardware, this is going to change. The ARD1939 protocol stack library is available as a free download.

However, before we dive into the programming of an SAE J1939 protocol stack, we need to cover some basics such as the Arduino hardware, some special programming topics, the CAN (Controller Area Network) protocol, and, naturally, the SAE J1939 standard.

1.1 The Arduino Uno And Mega 2560 plus CAN Shield

As I had mentioned earlier in this book, it is assumed that you have some basic knowledge of the Arduino Uno and the Mega 2560 itself, Arduino Sketches, and Arduino Shields.

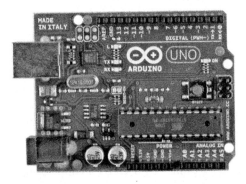

In order to develop and test the sample programs (sketches) as shown in this book, I initially used the Arduino Uno. The hardware consists of an open-source hardware board, usually designed around an 8-bit Atmel AVR microcontroller with 2 KB RAM (working memory), 32 KB Flash Memory (sketches) and 1 KB EEPROM (non-volatile).

These technical specifications are more than sufficient for basic prototyping of CAN and J1939 applications for a proof of concept. However, (and I will repeat this point) with growing demands for execution speed and extended functionality, the Arduino Uno may quickly reach its limits, specifically due to the sparse working memory of 2KB.

In order to overcome the Uno's memory restrictions (see also my notes on performance restrictions in a later chapter), I also used the Arduino Mega 2560, which, in turn, enabled me to run a full SAE J1939 protocol stack on the Arduino platform.

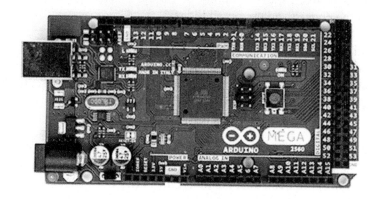

The Arduino Mega 2560 is a microcontroller board based on the ATmega2560 processor. Compared to the Arduino Uno, it comes (besides a great number of other advancements) with significantly improved memory size, specifically with 8 KB of SRAM compared to the Uno's 2 KB. In addition, it provides 256 KB Flash Memory (sketches) and 4 KB EEPROM (non-volatile).

In terms of programming, both versions, the Uno and the Mega 2560, are compatible.

Note: *All Arduino programs (sketches) as shown in this book were developed and tested with the Arduino Uno and Mega 2560. There is no guarantee that these programs will work "as is" on any other compatible system.*

It seems to be an obvious point, but many electronics enthusiasts initially don't realize that a network requires at least two nodes; otherwise you won't be able to establish a communication. Unless you have a real-world vehicle network (diesel engine or automobile) available for testing, you will need to buy a second Arduino with CAN shield or you can obtain a commercially available CAN gateway with Windows interface (i.e. a CAN analyzing and monitoring software).

The following image shows my network configuration with the Arduino Uno and Mega 2560 (I actually have several more CAN/J1939 nodes in order to simulate a full vehicle network).

For the network wiring I use regular telephone or speaker wires, which is absolutely acceptable under lab conditions. Just be aware that this kind of wiring may cause problems when connected to an actual vehicle.

For "real-world" cabling, I recommend browsing the Internet for SAE J1939 or ODB-II cables.

Also, in order to assure proper functionality, you should be aware that your network needs to utilize termination resistors as demonstrated in the following image (See also chapter *Introduction to Controller Area Network*).

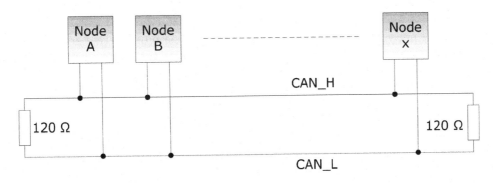

The rule of thumb is to use termination resistors at each end of the network, i.e. a maximum of two resistors. The CAN shields I will introduce in a later chapter have termination resistors on board, but they are not switchable.

When you work with more than two CAN/J1939 nodes (for instance, two Arduinos with CAN shield plus a CAN/J1939 gateway like I do), you need to be aware of the circumstance that you may have more than two termination resistors in the network, which may result in network disruptions. In all consequence, you may have to remove the resistor(s) (Refer to manufacturers' information on the location and/or removal of the termination resistors).

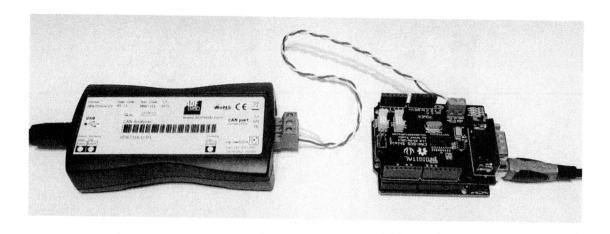

In order to test and verify the proper transmission and reception of CAN messages, I use the ADFweb CAN-to-USB gateway with its Windows interface.

Note: *In order to test a CAN/J1939 application, you need at least two CAN/J1939 nodes to establish a network communication. The second node can be another Arduino with CAN shield or (if the budget allows) another CAN/J1939 device with CAN/J1939 data monitoring capabilities.*

1.1.1 CAN Shield

For those who are not yet familiar with the SAE J1939 vehicle protocol (but are using this book to gain some more insights on the topic), J1939 uses Controller Area Network (CAN) controllers for the physical connection (See also chapter *A Brief Introduction to the SAE J1939 Protocol*). The Arduino Uno and the Mega 2560 do not, per default, come with an onboard CAN controller, and we need to use a CAN shield.

Since Controller Area Network (CAN) is predominantly targeted at industrial solutions (versus the vastly more popular USB for non-industrial use such as home and laboratory), there aren't too many choices available in the market.

Through some research (i.e. browsing) I found two very similar solutions, and they both work with the same CAN library (as explained in a later chapter). Both solutions use the Microchip MCP2515 CAN controller. Also, both solutions are distributed through worldwide online resources.

Note: *In the following I provide links to products that I used while writing this book. However, products and their capabilities may have changed since the publication of this book. It is highly recommended to check the manufacturers' websites (or those of distributors and representatives) for updated information.*

1.1.1.1 Microchip MCP2515 CAN Controller

Microchip Technology's MCP2515 is a stand-alone Controller Area Network (CAN) controller that implements the CAN specification, version 2.0B. It is capable of transmitting and receiving both standard and extended data and remote frames. The MCP2515 has two acceptance masks and six acceptance filters that are used to filter out unwanted messages, thereby reducing the host MCUs overhead. The MCP2515 interfaces with microcontrollers (MCUs) via an industry standard Serial Peripheral Interface (SPI).

The features include two receive buffers with prioritized message storage, six 29-bit filters, two 29-bit masks, and three transmit buffers with prioritization and abort features. *(Source: Microchip Datasheet)*

Note: *CAN specification 2.0B refers to the capability of using standard CAN frames with 11-bit message identifier plus the extended format with a 29-bit message ID (as required for J1939).*

To download the full MCP2515 datasheet log on to:
http://ww1.microchip.com/downloads/en/DeviceDoc/21801G.pdf

Both CAN shields as described in the following chapters also utilize the Microchip MCP2551 CAN transceiver, which converts the internal TTL signals to a differential voltage as demanded by the CAN standard.

To download the full MCP2551 datasheet log on to:
http://ww1.microchip.com/downloads/en/DeviceDoc/21667f.pdf

Note: *While the MCP2515 provides message filters and masks, their programming presented a challenge to a point of virtual uselessness (the original code delivered by Microchip is useless per default). The MCP2515 driver code as used in this book's Arduino projects does refer to message filtering, but the author of the code does not guarantee proper functionality. For that reason, all message filtering in the introduced Arduino projects is accomplished per software.*

1.1.1.2 Arduino CAN-Bus Shield by SK Pang electronics

This shield by SK Pang electronics provides the Arduino CAN-Bus capability. As explained previously, it uses the Microchip MCP2515 CAN controller with MCP2551 CAN transceiver. The CAN connection is realized via a standard 9-way sub-D, however the pin assignment for CAN_H, CAN_L is not according to standard.

Note: *In all truth, there is no mandatory standard for pin assignment, but the industry uses pins 2 (CAN_L) and 7 (CAN_H) as a virtual standard.*

I recommend using the on-board CAN_L and CAN_H contacts to solder the CAN cable directly to the board. The shield also comes with a uSD cardholder, a serial LCD connector, and a connector for an EM406 GPS module, making this shield suitable for data logging application.

Features

- CAN v2.0B up to 1 Mb/s
- High speed SPI Interface (10 MHz)
- Standard and extended data and remote frames
- CAN connection via standard 9-pin sub-D connector
- As an option, power can be supplied to the Arduino by sub-D via resettable fuse and reverse polarity protection.
- Socket for EM406 GPS module
- Micro SD card holder
- Connector for serial LCD
- Reset button
- Joystick control menu navigation control
- Two LED indicator

Notes

- No cables included
- Header pins are not included; they must be ordered separately
- Pin assignment for CAN_H, CAN_L not according to standard

All technical information regarding the use of the CAN controller, uSD cardholder, joystick, LEDs, etc. can be found on the company's wiki website at: https://code.google.com/p/skpang/

Ordering Information

To order the SK Pang electronics CAN shield, you can use the following resources (or browse for "Arduino CAN-BUS Shield" for further options):

Sparkfun - https://www.sparkfun.com/products/10039

SK Pang electronics - http://skpang.co.uk/catalog/arduino-canbus-shield-with-usd-card-holder-p-706.html

Note: *The version of the SK Pang CAN shield I used for this book worked fine with the Arduino Uno R3, but it failed when running it with the Arduino Mega 2560, but I had neither the time nor the resources to pinpoint the problem. It may well be that newer versions of the shield will work. It is recommended to check with the manufacturer in regards to compatibility with Arduino versions other than the Uno.*

1.1.1.3 CAN-BUS Shield by Seeed Studio

In terms of CAN capabilities, the shield by Seeek Studio provides the same functionality as the one by SK Pang electronics, however, it comes with a much lower price tag, because it does not have any additional components besides the CAN interface.

Note: *The Seeed Studio CAN Bus shield was my preferred choice during writing this book due to its compatibility between the two systems I primarily used, the Arduino Uno and Arduino Mega 2560. The Mega with its increased memory resources enabled me to run a full SAE J1939 protocol stack on the Arduino platform, which is not possible with the Uno.*

Over all, the device makes a solid impression, especially since the CAN connection is according to standard and in addition provides CAN connectivity through easily accessible terminals.

Features

- Implements CAN V2.0B at up to 1 Mb/s
- SPI Interface up to 10 MHz
- Standard (11 bit) and extended (29 bit) data and remote frames
- Two receive buffers with prioritized message storage
- Industrial standard 9 pin sub-D connector
- Two LED indicators

Notes

- No cables included

8

All technical information regarding the use of the CAN controller can be found on the company's wiki website at http://www.seeedstudio.com/wiki/CAN-BUS_Shield

Ordering Information

To order the Seeed Studio CAN shield, you can use the following resources (or browse for "Arduino CAN-BUS Shield" for further options):

Seeed Studio - http://www.seeedstudio.com/depot/CANBUS-shield-p-2256.html

<u>Note</u>: *The Seeed Studio CAN bus shield has been undergoing some hardware changes to become compatible with systems such as the Arduino Mega 2560. The version 1.0 will work with the Arduino Uno, while all higher versions also work with the Mega 2560. This will also affect the code of the Arduino projects, specifically the line "MCP_CAN CAN0(10);" in the main module selecting the CS pin. That line must change to "MCP_CAN CAN0(9);" for all CAN bus shield versions above 1.0. I have added a comment in the corresponding section of the code.*

1.1.2 Performance Restrictions

It is important to know that neither the Arduino Uno nor the Arduino Mega 2560, even though perfect for prototyping due to low price and ease of programming, are not, in their bare form, an industrial-strength solution, not only in terms of environmental specs (e.g. temperature range, etc.) but also in regards to execution speed and, specifically with the Uno, memory resources.

For instance, when it comes to CAN applications at 1 Mbit/sec and high data traffic, the Arduino Uno will reach its limits quickly. In addition, both versions, the Uno and the Mega 2560, are in their current form not suitable for a full CAN-to-USB gateway application due to speed restrictions of the on-board USB interface (See also next chapter).

The picture changes, however, when it comes to SAE J1939 applications, which require only a baud rate of 250 Kbit/sec. In all consequence, the Arduino Uno is well suited for SAE J1939 monitoring and simulation applications in regards to execution speed.

However, its limited memory resources, specifically the sparse 2k bytes of SRAM make it, unfortunately, impossible to implement and run a full-featured J1939 protocol stack. I will, nevertheless, present a solution (i.e. a limited version of the ARD1939 protocol stack) in the chapters to follow.

The problem is presented by the Transport Protocol (TP) according to SAE J1939/21, managing the transmission of data frames larger than the standard CAN 8 byte format. The TP supports up to 1785 bytes per message, and you will need at least twice that buffer size, since the SAE specification demands that a J1939 control application (CA) supports two sessions (BAM Session and RTS/CTS Session) concurrently. Even with extensive code tweaking, for instance, running only one concurrent session, there is no solution that would get you around using these 1785 bytes, and the remaining 263 bytes are easily exhausted.

Nevertheless, the Arduino Uno is very well suited for an implementation with no or limited TP support. In the case of the ARD1939 protocol stack (which will be introduced in a later chapter) the maximum data frame length is limited to 256 bytes, which is still more than useful for mere test and simulation purposes.

When it comes to the mere monitoring of J1939 vehicle data traffic (which does not require a J1939 protocol stack implementation), the Arduino Uno is just as suitable as any other, high-powered processor system.

In order to support a full SAE J1939 protocol stack implementation, I used the Arduino Mega 2560 due to its close compatibility with the Arduino Uno, specifically the operating voltage of 5 V. It presents another low-cost solution but with significantly increased memory resources (8kB of SRAM instead of 2kB).

Other systems, such as the Arduino Due, provide better CPU performance with significantly increased memory resources at roughly the same price as the Mega 2560.

However, we are also entering the grey zone of Arduino Shield compatibility. According to the Arduino website (arduino.cc), the Due is compatible with all Arduino shields that work at 3.3V. The reality is that the vast majority of shields currently work at 5V. The official Arduino website is not a great help, either. It only states that the Mega 2560 is designed to be compatible with most shields designed for the Uno, in all consequence a very vague statement.

In my case it turned out that, initially, neither one of the two CAN Bus Shields (Seeed Studio and SK Pang) were recognized by the Mega 2560 (using the same software that runs on the Uno). The trouble with the Arduino Mega 2560 and the Seeed Studio CAN shield was due to an incompatible SPI interface pin assignment, which Seeed Studio corrected with newer versions.

In case of the Due, the problem lies in the 3.3V operating voltage, which prohibits the use of shields that run at 5V. In addition, the Due has actually its own dedicated CAN interface, theoretically eliminating the use of an additional CAN shield. However, at the time of this writing, there was no programming interface (code) available. To worsen the situation, the CAN interface lacks the necessary CAN transceiver, making it useless for any CAN application, unless you buy (or create) an external breakout board.

And yes, there are many more Arduino options with better performance and increased memory resources, but digging through all possible hardware designs and testing the most promising scenarios can be an expensive and time-consuming project.

In the chapters to follow I will present several variations on how to use the Arduino Uno and Arduino Mega 2560 for SAE J1939 applications:

- **SAE J1939 Simulation and Monitoring Examples**: These programming samples come with virtually no restriction in regards to their functionality, and, as long as the code samples are being kept within a reasonable range, there should be no technical problem in terms of memory size.

- **Simple SAE J1939 to PC Gateway**: This application will work on both versions, Uno and Mega2560, since it does not involve a protocol stack implementation on the Arduino. The application simply passes 29-Bit CAN messages between the Arduino and the PC.

- **ARD1939 – Embedded J1939 Protocol Stack for Arduino**: ARD1939 supports the full SAE J1939 protocol stack including the Address Claim Procedure (SAE J1939/81) and the Transport Protocol (TP, SAE J1939/21), however, in case of the Arduino Uno the maximum message length is (due to memory size problems as explained earlier) 256 bytes. The version for the Arduino Mega 2560 comes with no restrictions in regards to message lengths, i.e. it does support the standard of up to 1785 bytes per message.

11

1.1.3 Serial Interface And CAN Timing Considerations

The hardware configuration as introduced in this book, namely the Arduino board plus CAN shield, suggests that it may be well suitable for a CAN-to-USB or J1939-to-USB gateway application. However, the Arduino's USB baud rate is limited to 115,200 baud, which is roughly half the speed of a 250 Kbit/sec J1939 connection, and roughly 11% of a full-blown 1Mbit/sec CAN application.

A reliable gateway application with the Arduino is only possible when the maximum CAN baud rate is not higher than 100 Kbit/sec, otherwise there is the risk of losing messages.

All this is under the assumption that the application is supposed to catch every single data frame in the network. One way to get around the timing problem is the use of message filters, i.e. the application allows only a set of application-specific message IDs or PGNs, respectively.

The reliability of an application using message filters depends heavily on the busload, i.e. the average frequency of CAN messages on the bus.

That being said, using the Arduino as a CAN-to-USB gateway comes with a great potential of losing messages. Even the use of large CAN message buffers does not guarantee that all messages are being received.

The situation is different with J1939 applications, because:

- The SAE J1939 standard dictates a 250 Kbit/sec baud rate.
- A typical J1939 application will only use a small subset of all available PGNs, which makes the use of message filters a logical choice.
- The highest message frequency in the SAE J1939/71 standard is 10 milliseconds.
- A busload of 60% is considered extremely high (100% would be continuous data flow).

It is therefore possible to create a full J1939-to-USB gateway with the Arduino, provided that some functionality such as effective message buffering is applied. Another measure would be to create interrupt service routines for receiving CAN and USB data.

Another aspect is the communication protocol between the Arduino's USB interface and the host system that receives the USB data. After all, in such a gateway application the CAN data is being converted into USB messages (and vice versa) and these messages must have a certain format. The protocol describes the conversion of these messages.

The more information is packaged into the protocol (for instance, stuffing bytes and checksum), the longer the message will be and the longer it takes to transmit it.

12

The average J1939 message takes roughly 540 microseconds to transmit. It will take the Arduino's USB port roughly 700 microseconds to transmit the same message in the best-case scenario (through stripping away CAN protocol overhead and ignoring any safety features such as a checksum). Any additions to the protocol (= protocol overhead) will result in transmission times in the range of 1,000 to 2,000 microseconds.

While all this results into a slower transmission of data frames by a factor of roughly 1.4 to almost 4, the Arduino will still be suitable for a basic gateway application.

Note: *Industrial-strength CAN-to-USB or J1939-to-USB gateways with their higher USB transmission rates in the Mbit/sec range use the luxury of extensive protocols that include, for instance, CAN frame time stamps.*

1.1.4 Connecting Arduino to a Real SAE J1939 Network

I have successfully tested my Arduino sketches against commercially available SAE J1939 devices with their monitoring and simulation software, but I did not have the luxury of connecting them to a real vehicle (diesel engine). Nevertheless, all sketches should work in the "real world."

If you do so, please read the description of all Arduino projects carefully. All sketches were created for mere demonstration and educational purposes, but their impact on a real vehicle network depends heavily on the application you are trying to accomplish. The mere monitoring of J1939 data frames should post no problem whatsoever, but writing data into the network requires detailed knowledge of the network and its components. I have tried to address any such aspect in the individual projects.

In order to connect your Arduino to a real vehicle network, you will need to provide the proper wiring and connectors. In the following, I will explain connections as they are used in the industry, but my description can only serve as a coarse educational example. Please refer to the SAE J1939 Standards Collection (SAE J1939/1x and SAE J1939/2x documents) for proper information.

1.1.4.1 SAE J1939/13 Off-Board Diagnostic Connector

J1939/13 defines a standard connector for diagnostic purpose. It does allow access to the vehicle communication links. The connector is a Deutsch HD10 - 9 – 1939 (9 pins, round connector).

According to the official document, *SAE J1939/13 Off-Board Diagnostics Connector*, the connector supports both the twisted shielded pair media (as defined in SAE J1939/11) as well as the twisted unshielded quad media (as defined by ISO 11783-2). The designations of the individual signal wires are according to the CAN Standard CAN_H and CAN_L. For SAE J1939/11, a third connection for the termination of the shield is denoted by CAN_SHLD.

The pin assignment is as follows:

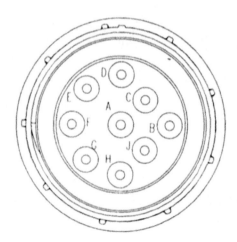

Pin A – Battery (-)
Pin B – Battery (+)
Pin C – CAN_H
Pin D – CAN_L
Pin E – CAN_SHLD
Pin F – SAE J1708 (+)
Pin G – SAE J1708 (-)
Pin H – Proprietary OEM Use or Implement Bus CAN_H
Pin J - Proprietary OEM Use or Implement Bus CAN_L

For more detailed information on the connector and its wiring, please refer to the official SAE document.

1.2 Arduino Programming (Sketches)

The implementation of either one of the introduced CAN-BUS Shields and the corresponding CAN sketches went surprisingly smooth when paired with the right library software.

Note: *All Arduino CAN/J1939 sketches as explained in the following include the MPC2515 library plus the interface between the CAN and SAE J1939 layers. If you are eager to jump into the J1939 programming, you may skip this chapter and use the information when you start your own CAN/J1939 projects.*

I found several source code examples for accessing the MCP2515 CAN controller, but most of them didn't even pass the initial quality control phase (I read the code first before I use it). One of the quality criteria was the support for 29-bit CAN message identifiers (CAN 2.0B Compatibility), which is mandatory when it comes to implementing the SAE J1939 vehicle network protocol. Some software samples I found were just literally "samples" and they left ample room for guessing games.

I was most pleased by the MPC2515 Library by Cory Fowler, which can be found at https://github.com/coryjfowler/MCP2515_lib.

This library is compatible with any shield or CAN interface that uses the MCP2515 CAN protocol controller.

1.2.1 The MCP2515 Library

As with any serial networking controller, the essential function calls are:

1. Initialization
2. Read Data
3. Write Data
4. Check Status

In case of the MCP2515 library, these functions are represented by:

1. Initialization: CAN0.begin
2. Read Data: CAN0.readMsgBuf
   incl. CAN0.checkReceive, CAN0.getCanId
3. Write Data: CAN0.sendMsgBuf
4. Check Status: CAN0.checkError

1.2.1.1 Function Calls

Function: **CAN0.begin**
Purpose: Initializes the CAN controller and sets the speed (baud rate)
Parameter: CAN_5KPS ... CAN_1000KPS (See mcp_can_dfs.h)
Return Code: CAN_OK = Initialization okay
CAN_FAILINIT = Initialization failed

Function: **CAN0.checkReceive**
Purpose: Check if message was received
Parameter: None
Return Code: CAN_MSGAVAIL = Message available
CAN_NOMSG = No message

Function: **CAN0.readMsgBuf**
Purpose: Read the message buffer
Parameter: nMsgLen returns the message length (number of data bytes)
nMsgBuffer returns the actual message
Return Code: None

Function: **CAN0.getCANId**
Purpose: Retrieves the ID of the received message
Parameter: None
Return Code: m_nID = Message ID

Function:	**CAN0. sendMsgBuf**
Purpose:	Send a message buffer
Parameter:	id = Message ID
	ext = CAN_STDID (11-bit ID) or CAN_EXTID (29-bit ID)
	len = Number of data bytes (0…8)
	buf = Message buffer
Return Code:	None

Function:	**CAN0.checkError**
Purpose:	Checks CAN controller for errors
Parameter:	None
Return Code:	CAN_OK = Status okay
	CAN_CTRLERROR = Error

There are further functions, among others, for message filtering and settings masks, and they are worth being checked out for more sophisticated functions, but they are not necessary for simple CAN communication tasks.

The implementation of the MPC2515 library is fairly easy: Open Arduino, create a new file, then use the menu items *Sketch->Add File...* to include the following files to the project:

- mcp_can.cpp
- mcp_can.h
- mcp_can_dfs.h

In the Arduino project file add the following on top:

```
#include "mcp_can.h"
#include <SPI.h>
MCP_CAN CAN0(10);
```

Let me repeat two important points regarding the MCP2515 here:

- The Seeed Studio CAN bus shield has been undergoing some hardware changes to become compatible with systems such as the Arduino Mega 2560. The version 1.0 will work with the Arduino Uno, while all higher versions also work with the Mega 2560. This will also affect the code of the Arduino projects, specifically the line "MCP_CAN CAN0(10);" in the main module selecting the CS pin. That line must change to "MCP_CAN CAN0(9);" for all CAN bus shield versions above 1.0.

- While the MCP2515 provides message filters and masks, their programming presented a challenge to a point of virtual uselessness (the original code delivered by Microchip is useless per default). The MCP2515 driver code as used in this book's Arduino projects does refer to message filtering, but the author of the code does not guarantee proper functionality. For that reason, all message filtering in the introduced Arduino projects is accomplished per software.

1.2.1.2 The CAN Interface

While the code as introduced in the previous chapter was well designed and thus is highly efficient, I inserted yet another software layer between the CAN interface and the ARD1939 protocol stack.

I wrote the ARD1939 source code in plain C (not C++) to assure the highest level of compatibility with other embedded systems and their compilers (yes, the code already runs under an ARM system, and I will publish another book about it). Naturally, systems other than the Arduino Uno or Mega 2560 require different CAN drivers, and it would be quite a cumbersome endeavor to find and replace all CAN function calls within the protocol.

For that reason, I created some "generic" function calls that are being used by the protocol stack. For future implementations, I only need to rewrite the **can.cpp** module.

Note: *In addition to the MCP2515 function calls, I added a CAN message ring buffer. The MCP2515's message buffer is limited, and with higher busload (high message frequency) you will lose messages.*

These function calls are yet again based on the basic structure of each serial communications program:

1. Initialization
2. Read Data
3. Write Data
4. Check Status

These functions are represented by:

1. Initialization: canInit()
2. Read Data: canReceive(…)
3. Write Data: canTransmit(…)
4. Check Status: canCheckError()

Note: *For highest compatibility between compilers, I refrained from using complex structures and their pointers. The KISS principle (Keep It Simple Stupid!) makes portations an easy task and it increases code readability. All that comes with virtually non-existing performance restraints.*

1.2.2 Special Programming Topics

Each programmer has its own coding style, and I am most certainly no exception. Nevertheless, code should always be written with the following aspects in mind:

- High level of readability, including plenty of comments (you will have trouble decoding your own programming only weeks after you looked at it the last time)
- Easy debugging (readability helps here, too)
- Possible memory restraints (applies especially to the Arduino Uno)
- Keep It Simple Stupid (it doesn't make any sense spending valuable time analyzing/documenting overly complex structures, enum arrays, etc. unless they provide great advantages in respect to performance)

With my particular programming style, I try to follow these rules, especially under the aspect that I am sharing my code with a large community. If the code isn't simple and intuitive enough for my readers to figure it out without asking any questions, it usually will be rejected.

1.2.2.1 Hungarian Notation

Hungarian notation is an identifier naming convention in computer programming, in which the name of a variable or function indicates its type or intended use. Hungarian notation was designed to be language-independent, and found its first major use with the BCPL programming language. Because BCPL has no data types other than the machine word, nothing in the language itself helps a programmer remember variables' types. Hungarian notation aims to remedy this by providing the programmer with explicit knowledge of each variable's data type.

In Hungarian notation, a variable name starts with a group of lower-case letters, which are mnemonics for the type or purpose of that variable, followed by whatever name the programmer has chosen; this last part is sometimes distinguished as the given name.

The original Hungarian notation, which would now be called Apps Hungarian, was invented by Charles Simonyi, a programmer who worked at Xerox PARC circa 1972–1981, and who later became Chief Architect at Microsoft. It may have been derived from the earlier principle of using the first letter of a variable name to set its type — for example, variables whose names started with letters I through N in FORTRAN are integers by default. *(Source: Wikipedia.org)*

19

In my programming, without following the "official" Hungarian Notation to the last detail, I use descriptive variable names to improve the code's readability and they are preceded by one or two lower-case letters indicating the type of the variable.

For instance, I use *n* for integer and *s* for string. Just to list a few examples:

```
bool bCANDataReceived;
byte cSerialData;
int nJ1939Status;
char sJ1939Application[20];
```

I will refrain from providing an "official" list of my naming convention, because it does not exist. However, when you look at the programs I provide, you will get the idea (without the need of adapting the style for your applications). My mere intention is providing educational-style information based on great readability.

1.2.2.2 Debugging Code with Macros

While one of the Arduino Uno's strengths is its ease of programming embedded solutions, it can become increasingly frustrating when it comes to debugging the code. The Arduino IDE provides only very limited, if non-existing, debugging capabilities. The only solution to the problem is adding code that converts a variable to a string and display it on the Arduino's Serial Monitor, a task that can be cumbersome and time-consuming.

Other, professional (and much more costly) programming environments will allow you to set breakpoints and display the value of variables, even arrays. Until the time comes where the Arduino IDE provides such features, let's debug our code with the help of a few so-called *C Preprocessor Macros*.

*The **C preprocessor** modifies a source code file before handing it over to the compiler. You're most likely used to using the preprocessor to include files directly into other files, or #define constants, but the preprocessor can also be used to create "inlined" code using macros expanded at compile time and to prevent code from being compiled twice.*

*The other major use of the preprocessor is to define **macros**. The advantage of a macro is that it can be type-neutral (this can also be a disadvantage, of course), and it's inlined directly into the code, so there isn't any function call overhead.*

Source: http://www.cprogramming.com/tutorial/cpreprocessor.html.

Using C Preprocessor Macros, I developed the following instructions to debug my Arduino applications:

- **DEBUG_INIT()** – Defines a string variable for debugging. This function is mandatory to initiate debugging, and it should be placed at the beginning of the code module (file) that you want to debug.
- **DEBUG_PRINTHEX(T, v)** – Prints a variable in hexadecimal format with preceding Text.
- **DEBUG_PRINTDEC(T, v)** - Prints a variable in decimal format with preceding Text.
- **DEBUG_PRINTARRAYHEX(T, a, l)** – Prints an array of variables with a certain length in hexadecimal format with preceding Text.
- **DEBUG_PRINTARRAYDEC(T, a, l)** – Prints an array of variables with a certain length in decimal format with preceding Text.
- **DEBUG_HALT()** – Stops the program until user submits a keystroke through the Serial Monitor.
  <u>Note</u>: The macro is reading one character at a time, i.e. if you submit more than one character, the program will continue until all characters have been read. This can, in certain cases, be a helpful feature.

The variables used for the macros are:

- **T** – Text to be displayed with the variable (e.g. name of the variable)
- **v** – The actual variable; can be of any type (e.g. int, byte, float, etc.)
- **a** – Pointer to an array of variables
- **l** – Length of the array

The DEBUG instructions, stored in a debug.h file, are part of all Arduino code projects in this book (See the actual code in the appendix).

The result of the debug prints is displayed on the Arduino's Serial Monitor. The following code represents a program sample that demonstrates the use of the macros:

```
DEBUG_INIT()

// Call the J1939 protocol stack
nJ1939Status = j1939Operate(&nMsgID, &lPGN, &pMsg[0], &nMsgLen,
                            &nDestAddr, &nSrcAddr, &nPriority);

DEBUG_PRINTHEX("nDestAddr = ", nDestAddr)
DEBUG_PRINTDEC("nDestAddr = ", nDestAddr)

DEBUG_PRINTARRAYHEX("pMsg = ", pMsg, nMsgLen)
DEBUG_PRINTARRAYDEC("pMsg = ", pMsg, nMsgLen)

DEBUG_HALT()
```

2. A Brief Introduction to the SAE J1939 Protocol

The Society of Automotive Engineers (SAE) Truck and Bus Control and Communications Subcommittee has developed a family of standards concerning the design and use of devices that transmit electronic signals and control information among vehicle components. SAE J1939 and its companion documents have quickly become the accepted industry standard and the vehicle network of choice for off-highway machines in applications such as construction, material handling, mass transportation, forestry machines, agricultural machinery, maritime and military applications.

Note: *It is fair to say that the SAE J1939 vehicle protocol is primarily used on diesel engines, which covers all previously mentioned applications.*

J1939 is a higher-layer protocol based on Controller Area Network (CAN). It provides serial data communications between microprocessor systems (also called Electronic Control Units - ECU) in any kind of heavy-duty vehicles. The messages exchanged between these units can be data such as vehicle road speed, torque control message from the transmission to the engine, oil temperature, and many more.

Note: *Even though being around for many years, the SAE J1939 protocol is still gaining popularity, especially in view of the increased use of fleet management systems, which, unavoidably, will need data from the vehicle network, for instance, to calculate maintenance cycles. Fleet management is also tightly associated with the Internet of Things (IoT), and transportation is considered one of the fastest growing markets for IoT.*

2.1 Introduction to Controller Area Network

Controller Area Network (CAN) is a serial network technology that was originally designed for the automotive industry, especially for European cars, but has also become a popular bus in industrial automation as well as other applications. The CAN bus is primarily used in embedded systems, and as its name implies, is a network technology that provides fast communication among microcontrollers up to real-time requirements, eliminating the need for the much more expensive and complex technology of a Dual-Ported RAM.

CAN is a two-wire, half duplex, high-speed network system, that is far superior to conventional serial technologies such as RS232 in regards to functionality and reliability and yet CAN implementations are more cost effective.

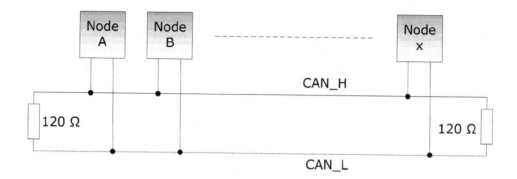

While, for instance, TCP/IP is designed for the transport of large data amounts, CAN is designed for real-time requirements and with its 1 MBit/sec baud rate can easily beat a 100 MBit/sec TCP/IP connection when it comes to short reaction times, timely error detection, quick error recovery and error repair.

CAN networks can be used as an embedded communication system for microcontrollers as well as an open communication system for intelligent devices. Some users, for example in the field of medical engineering, opted for CAN because they have to meet particularly stringent safety requirements. Similar requirements had to be considered by manufacturers of other equipment with very high safety or reliability requirements (e.g. robots, lifts and transportation systems).

The greatest advantage of Controller Area Network lies in the reduced amount of wiring combined with an ingenious prevention of message collision (meaning no data will be lost during message transmission).

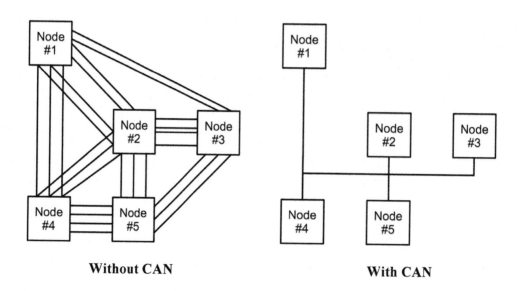

Without CAN   **With CAN**

The following shows a need-to-know overview of CAN's technical characteristics.

Controller Area Network

- Is a serial networking technology for embedded solutions.

- Needs only two wires named CAN_H and CAN_L.

- Operates at data rates of up to 1 Megabit per second.

- Supports a maximum of 8 bytes per message frame.

- Does not support node IDs, only message IDs. One application can support multiple Message IDs.

- Supports message priority, i.e. the lower the message ID the higher its priority.

- Supports two message ID lengths, 11-bit (standard) and 29-bit (extended).

- Does not experience message collisions (as they can occur under other serial technologies).

- Is not demanding in terms of cable requirements. Twisted-pair wiring is sufficient.

Note: *For more detailed information on CAN, please refer to the official CiA/Bosch specification or to "A Comprehensible Guide to Controller Area Network" as mentioned in the literature appendix of this book.*

2.2 The SAE J1939 Higher-Layer Protocol

Even though extremely effective in automobiles and small, embedded applications, CAN alone is not suitable for projects that require a minimum of network management and messages with more than eight data bytes.

As a consequence, higher layer protocols (additional software on top of the CAN physical layer) such as SAE J1939 for vehicles were designed to provide an improved networking technology that support messages of unlimited length and allow network management, which includes the use of node IDs (CAN supports only message IDs where one node can manage multiple message IDs).

SAE J1939

- Is a standard developed by the Society of Automotive Engineers (SAE)
- Defines communication for vehicle networks (trucks, buses, agricultural equipment, etc.)
- Is a Higher-Layer Protocol using CAN as the physical layer
- Uses shielded twisted pair wire
- Applies a maximum network length of 40 meters (~120 ft.)
- Applies a standard baud rate of 250 Kbit/sec
- Allows a maximum of 30 nodes (ECUs) in a network
- Allows a maximum of 253 controller applications (CA) where one ECU can manage several CAs
- Supports peer-to-peer and broadcast communication
- Supports message lengths up to 1785 bytes
- Defines a set of Parameter Group Numbers (PGNs, predefined vehicle parameters)
- Supports network management (includes node IDs and an address claiming procedure)

Compared to other, function-driven protocols such as CANopen and DeviceNet, SAE J1939 is primarily data-driven. In fact, J1939 provides a far better data bandwidth than any of these automation protocols.

J1939 data packets contain the actual data and a header, which contains an index called Parameter Group Number (PGN). A PGN identifies a message's function and associated data (a more detailed description of PGNs and essential protocol functions are included in the description of the ARD1939 protocol stack later in this book). J1939 attempts to define standard PGNs to encompass a wide range of automotive, agricultural, marine and off-road vehicle purposes.

Note: *For more detailed information on the SAE J1939 Standards Collection, please refer to the official SAE documentation or to "A Comprehensible Guide to J1939" as mentioned in the literature appendix of this book.*

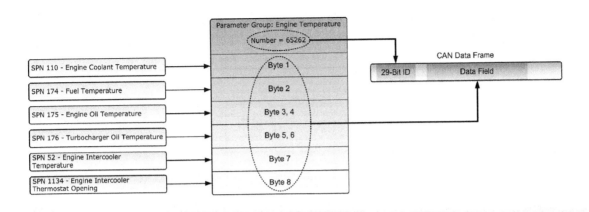

The actual data in the data field is described by the SPNs.

Note: *You will need a copy of the SAE J1939/71 standard for a detailed description of all PGNs and SPNs. You may find some of them by browsing the Internet, but be aware that there is no complete online reference. A description of all PGNs is out of the scope of this book.*

2.3 Parameter Group Numbers (PGN)

SAE J1939 is a very ingeniously designed protocol that takes a resourceful advantage of the CAN 29-Bit message identifier. Rather than relying on a myriad of protocol functions, SAE J1939 uses predefined parameter tables, which keeps the actual protocol on a comprehensible level.

Parameters groups are, for instance, engine temperature, which includes coolant temperature, fuel temperature, oil temperature, etc. Parameter Groups and their numbers (PGN) are listed in SAE J1939 (roughly 300 pages) and defined in SAE J1939/71, a document containing roughly 800 pages filled with parameter group definitions plus suspect parameter numbers (SPN). In addition, it is possible to use manufacturer-specific parameter groups.

From a programming aspect, it is important to know that the combination of PGN, message priority, and the application's node (source) address represents the 29-Bit message identifier of the CAN message as demonstrated in the following image.

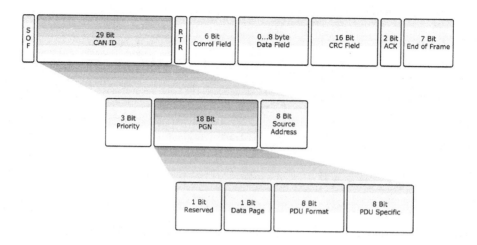

Note: *For more detailed information on the J1939 message format, please refer to the official SAE documentation or my book "A Comprehensible Guide to J1939." However, the definition of, for instance, "PDU Format" and "PDU Specific," come with a great potential of confusing the J1939 novice, i.e. the educational value is questionable.*

2.3.1 Parameter Groups

Parameter Groups contain information on parameter assignments within the 8-byte CAN data field of each message as well as repetition rate and priority.

The following is an example of a parameter group definition as listed in SAE J1939/71:

- **PGN 65262** - Engine Temperature
    - Transmission Rate: 1 sec
    - Data Length: 8 bytes
    - Default Priority: 6
    - PG Number: 65262 (FEEEhex)

- **Description of Data**
    - Byte 1: Engine Coolant Temperature – SPN 110
    - Byte 2: Fuel Temperature – SPN 174
    - Byte 3, 4: Engine Oil Temperature – SPN 175
    - Byte 5, 6: Turbocharger Oil Temperature- SPN 176
    - Byte 7: Engine Intercooler Temperature - SPN 52
    - Byte 8: Engine Intercooler Thermostat Opening - SPN 1134

2.3.2 Suspect Parameter Numbers (SPN)

A Suspect Parameter Number (SPN) is a number assigned by the SAE to a specific parameter within a parameter group. It describes the parameter in detail by providing the following information:

- Data Length in bytes
- Data Type
- Resolution
- Offset
- Range
- Reference Tag (Label)

SPNs that share common characteristics are grouped into Parameter Groups (PG) and they will be transmitted throughout the network using the Parameter Group Number (PGN).

To follow up on the previous example (PGN 65262), the parameter *Engine Coolant Temperature* is described by SPN 110 in the following way:

- **SPN 110** - Engine Coolant Temperature
  Temperature of liquid engine cooling system
  - Data Length: 1 Byte
  - Resolution: 1 deg C / Bit
  - Offset: -40 deg C
  - Data Range: -40 to 210 deg C
  - Type: Measured
  - Reference: PGN 65262

2.3.3 PGN Range

Program Parameter Numbers (PGNs) are in a range of:

- 0x0000 – 0xEEFF: 239 Peer-to-Peer messages defined by SAE
- 0xEF00 – 0xEFFF: 1 Peer-to-Peer message for proprietary use
- 0xF000 – 0xFEFF: 3840 Broadcast messages defined by SAE
- 0xFF00 – 0xFFFF: 256 Broadcast messages for proprietary use

Note: *The SAE J1939 allows two types of messages, peer-to-peer (= direct node communication) and broadcast. Broadcast messages (their message ID includes the address of the sending node) are distributed to all nodes and the nodes decide whether to use it or not. Peer-to-Peer messages use a message ID that includes the transmitter and receiver address. Node addresses are always 8 bits long.*

2.4 The Two Elements of the SAE J1939 Protocol Stack

Any SAE J1939 hardware must support SAE J1939/1x and SAE J1939/21, otherwise they're useless, because these standards describe the CAN bus physical layer and the basic protocol features. That part is sufficiently covered (i.e. for demonstration purposes) by our CAN shield, while the cabling and connectors may not necessarily J1939-compliant.

A fully functional SAE J1939 protocol software must support one mandatory element, SAE J1939/81 (address claim process) and, if needed, SAE J1939/21 (transport of up to 1785 bytes per message).

Note: *The implementation of the Transport Protocol (TP), i.e. the transport of up to 1785 data bytes in a message, is highly application-specific. Some ECUs and their control applications (CAs) will need it, some won't.*

In the following, I will cover the basics of the SAE J1939/21 Transport Protocol (TP) and the Address Claim Procedure according to SAE J1939/81. The information as presented is in reference to help understanding the ARD1939 protocol stack.

Note: *For a more hand-on approach to the SAE J1939/21 and SAE J1939/81 requirements, please refer to chapter "Proof of Concept." The tests accomplished in that chapter explain the actual SAE J1939 protocol in more detail than any other textbook or manual.*

2.4.1 SAE J1939/21 - Transport Protocol (TP)

Even though extremely effective in passenger cars and small industrial applications, CAN alone was not suitable to meet the requirements of truck and bus communications, especially since its communication between devices is limited to only 8 bytes per message. However, it is possible to extend the size of a CAN message by implementing additional software, i.e. so-called higher layer protocols. J1939 is such a higher layer protocol and it supports up to 1785 bytes per message.

In order to support a size of more than 8 bytes the message needs to be packaged into a sequence of 8 byte size messages. Consequently, the receiver of such a multi-packet message must re-assemble the data. Such functions are defined as Transport Protocol (TP) Functions and they are described in SAE J1939/21.

In order to package CAN messages into a sequence of up to 1785 messages (as well as to re-assemble the CAN frames into one data package) the J1939 Transport Protocol defines the following:

- Each multi-packet message is being transmitted by using a dedicated Data Transfer PGN (60160, TP.DT = Transport Protocol Data Transfer), i.e. all message packets will have the same ID.

- The flow control is managed by another dedicated PGN (60146, TP.CM = Transport Protocol Communication Management).

- The message length must always be 8 bytes (DLC = 8).

- The first byte in the data field contains a sequence number that ranges from 1 to 255.

- The remaining 7 bytes are filled with the data of the original long (> 8 bytes) message.

- All unused data bytes in the last package are being set to 0xFF.

The method of using a sequence number plus the remaining seven data bytes yields a total of (255 packages times 7 bytes/package) 1785 bytes per multi-packet message.

The following images demonstrate the use of the CAN data field including the sequence number and the packaging of multiple CAN messages.

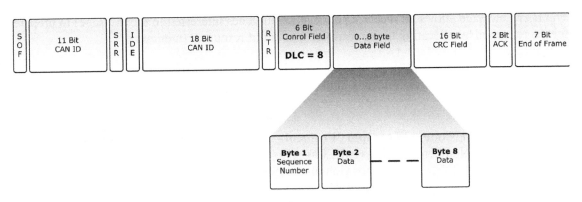

CAN Data Frame with Sequence Number

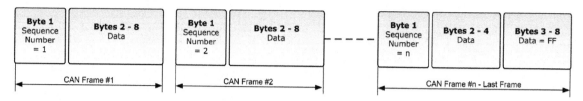

Example of Multi-Package Sequence

As it is the case with regular 8-byte CAN messages, multi-packet messages (i.e. messages longer than 8 bytes) can be transmitted as a broadcast message (BAM = Broadcast Announce message) or peer-to-peer (RTS/CTS = Request to Send / Clear to Send).

Note: *The SAE J1939/21 Standard requires that an ECU supports one BAM and one RTS/CTS session simultaneously.*

2.4.1.1 Multi-Packet Broadcast (BAM Session)

In order to broadcast a multi-packet message, a node must first send the *Broadcast Announce Message (BAM)*, which contains the following components:

- Parameter Group Number (PGN) of the multi-packet message
- Size of the multi-packet message
- Number of packages

The BAM message allows all receiving nodes (i.e. all nodes interested in the message) to prepare for the reception by allocating the appropriate amount of resources (memory).

The *Broadcast Announce Message* (BAM) is embedded in the Transport Protocol – Connection Management (TP.CM) PGN 60416 and the actual data transfer is handled by using the Data Transfer PGN 60160.

Parameter Group Name	Transport Protocol – Connection Management (TP.CM)
Parameter Group Number	60416 (0xEC00)
Definition	Used for Communication Management flow-control (e.g. Broadcast Announce Message).
Transmission Rate	According to the Parameter Group Number to be transferred
Data Length	8 bytes
Extended Data Page (R)	0
Data Page	0
PDU Format	236
PDU Specific	Destination Address (= 255 for broadcast)

Default Priority	7
Data Description	(For Broadcast Announce Message only)
Byte	1 - Control Byte = 32
	2,3 – Message Size (Number of bytes; LSB first, then MSB)
	4 – Total number of packages
	5 – Reserved (should be filled with 0xFF)
	6-8 – Parameter Group Number of the multi-packet message (6=LSB, 8=MSB)

Parameter Group Name	**Transport Protocol – Data Transfer (TP.DT)**
Parameter Group Number	60160 (0xEB00)
Definition	Data Transfer of Multi-Packet Messages
Transmission Rate	According to the Parameter Group Number to be transferred
Data Length	8 bytes
Extended Data Page (R)	0
Data Page	0
PDU Format	235
PDU Specific	Destination Address
Default Priority	7
Data Description	
Byte	1 – Sequence Number (1 to 255)
	2-8 - Data

The last packet of a multi-packet PGN may require less than eight data bytes. All unused data bytes in the last package are being set to 0xFF.

The transport of Multi-Packet Broadcast messages is not regulated by any flow-control functions and thus it is necessary to define timing requirements between the sending of a *Broadcast Announce Message* (BAM) and the Data Transfer PGN. The following picture demonstrates the message sequence and timing requirements for a broadcasted multi-packet message.

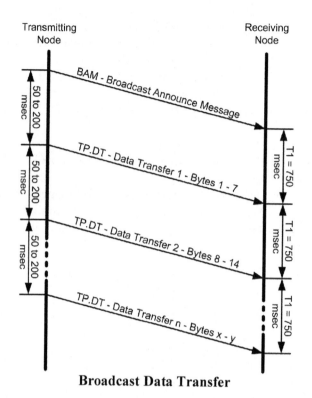

Broadcast Data Transfer

As is demonstrated in the previous image:

- The message packet frequency will be between 50 to 200 msec.
- A timeout will occur when a time of greater than 750 ms (T1) elapsed between two message packets when more packets were expected.

Timeouts will cause a connection closure.

A connection is considered closed when:

- The sender of a data message sends the last Data Transfer package.
- A timeout occurs.

Note: *The SAE J1939/21 document is not very specific regarding the time range of 50 to 200 milliseconds, i.e. how the range was defined and which time should be applied under which conditions.*

First of all, a minimum of 50 milliseconds creates ample time for other message to go through the network. Using the lowest message priority assures that higher-priority messages still reach the bus in a timely manner. In other words, the combination of a long time and low priority guarantees to a certain degree that long data messages do not disrupt the regular data traffic.

The range of 50 to 200 milliseconds, however, still remains a mystery. It may be based on the technology when SAE J1939 was introduced, i.e. there may have been ECUs with a reaction time of up to 200 milliseconds. Another, more plausible explanation may be that the time can be chosen due to application needs, meaning 200 milliseconds are less intrusive to the regular data traffic than 50 milliseconds.

2.4.1.2 Multi-Packet Peer-to-Peer (RTS/CTS Session)

The communication of destination specific (peer-to-peer) multi-packet messages is subject to flow-control.

1. **Connection Initialization** – The sender of a message transmits a *Request to Send* message. The receiving node responds with either a *Clear to Send* message or a *Connection Abort* message in case it decides not to establish the connection. A *Connection Abort* as a response to a *Request to Send* message is preferred over a timeout by the connection initiator. The *Clear to Send* message contains the number of packets the receiver is expecting plus the expected sequence number.

2. **Data Transfer** – The sender transmits the Data Transfer PGN after receiving the *Clear to Send* message. Data transfer can be interrupted/stopped by a *Connection Abort* message.

3. **Connection Closure** – The receiver of the message, upon reception of the last message packet, sends an *End of Message ACK* (acknowledgement) message, provided there were no errors during the transmission. Any node, sender or receiver, can send a *Connection Abort* message. The reason of aborting a connection can be a timeout.

A reliable flow-control will must also include timeouts in order to assure proper network function, and SAE J1939/21 defines a number of timeouts.

T_r = 200 ms - Response Time
T_h = 500 ms - Holding Time
T1 = 750 ms
T2 = 1250 ms
T3 = 1250 ms
T4 = 1050 ms

Scenarios for timeout control are:

- A node (regardless whether the node is the receiver or sender of the data message) does not respond within 200 ms (T_r) to a data or flow control message.

- If a receiving node needs (for any reason) to delay the transmission of data it can send a *Clear to Send* message where the number of packages is set to zero. In cases where the flow must be delayed for a certain time the receiver of a message must repeat the transmission of the *Clear to Send* message every 0.5 seconds **(Th)** to maintain an open connection with the sender of the message. As soon as the receiver is ready to receive the message it must send a regular *Clear to Send* message.

- A time of greater than **T1** elapsed between two message packets when more packets were expected.

- A time greater than **T2** elapsed after a *Clear to Send* message without receiving data from the sender of the data.

- A time greater than **T3** elapsed after the last transmitted data packet without receiving a *Clear to Send* or *End of Message Acknowledgment* (ACK) message.

- A time greater than **T4** elapsed after sending a Clear to Send message to delay data transmission without sending another Clear to Send message

Any timeout condition will consequently cause a connection closure.

Note: *What SAE J1939/21 fails to mention is that the "Clear to Send" message can be send by the receiver of the data message at any time, either immediately after the reception of a "Request to Send" message or after reception of a data packet, meaning any time during the data transfer.*

Other reasons for connection closure are:

- The sender of a data message sends the last Data Transfer package.

- The receiver of a data message receives the last Data Transfer package and a T1 timeout occurs.

- The sender of a data message receives an *End of Message ACK* message.

- Reception of *Connection Abort* message.

The flow control messages, such as *Request to Send, Clear to Send*, etc. are embedded in the Transport Protocol – Connection Management (TP.CM) PGN 60416 and the actual data transfer is handled by using the Data Transfer PGN 60160.

Parameter Group Name	**Transport Protocol – Connection Management (TP.CM)**
Parameter Group Number	60416 (0x00EC00)
Definition	Used for Communication Management flow-control (e.g. *Request to Send*, *Clear to Send*, etc.).
Transmission Rate	According to the Parameter Group Number to be transferred
Data Length	8 bytes
Extended Data Page (R)	0
Data Page	0
PDU Format	236
PDU Specific	Destination Address (= 255 for broadcast)
Default Priority	7
Data Description	Depending on content of Control Byte – See following description.

TP.CM_RTS

Connection Mode Request to Send
1 - Control Byte = 16
2,3 – Message Size (Number of bytes)
4 – Total number of packets
5 – Max. number of packets in response to CTS. No limit when filled with 0xFF.
6-8 – Parameter Group Number of the multi-packet message (6=LSB, 8=MSB)

TP.CM_CTS

Connection Mode Clear to Send
1 - Control Byte = 17
2 - Total number of packets (should not exceed byte 5 in RTS)
3 – Next packet number
4,5 – Reserved (should be filled with 0xFF)
6-8 – Parameter Group Number of the multi-packet message (6=LSB, 8=MSB)

TP.CM_EndOfMsgACK

End of Message Acknowledgment
1 - Control Byte = 19
2,3 – Message Size (Number of bytes)
4 – Total number of packages
5 – Reserved (should be filled with 0xFF)
6-8 – Parameter Group Number of the multi-packet message (6=LSB, 8=MSB)

TP.Conn_Abort

Connection Abort
1 - Control Byte = 255
2 – Connection Abort Reason (See following description)
3-5 – Reserved (should be filled with 0xFF)
6-8 – Parameter Group Number of the multi-packet message (6=LSB, 8=MSB)

Control Byte = 32 is reserved for *Broadcast Announce Message*. Control Bytes 0-15, 18, 20-31, 33-254 are reserved by the SAE.

The Connection Abort Reasons can be:

1 – Node is already engaged in another session and cannot maintain another connection.

2 – Node is lacking the necessary resources.

3 – A timeout occurred.

4...250 - Reserved by SAE.

251...255 – Per J1939/71 definitions (Unfortunately, the SAE J1939 Standards Collection offers no further explanations.)

The actual data transfer is handled by using the Data Transfer PGN 60160.

Parameter Group Name	Transport Protocol – Data Transfer (TP.DT)
Parameter Group Number	60160 (0x00EB00)
Definition	Data Transfer of Multi-Packet Messages
Transmission Rate	According to the Parameter Group Number to be transferred
Data Length	8 bytes
Extended Data Page (R)	0
Data Page	0
PDU Format	235
PDU Specific	Destination Address
Default Priority	7
Data Description	
Byte	1 – Sequence Number (1 to 255)
	2-8 – Data

The last packet of a multi-packet PGN may require less than eight data bytes. All unused data bytes in the last package are being set to 0xFF.

The following image demonstrates the basic data flow between the transmitting and receiving mode during an RTS/CTS session (i.e. peer-to-peer communication).

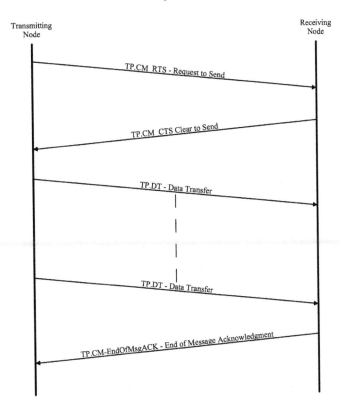

Peer-to-Peer data Transfer

Note: *I refrained from inserting the various timeouts that can occur during an RTS/CTS session. Each individual timeout scenario would require its own flow chart, while the functionality was already sufficiently described.*

2.4.2 SAE J1939/81 - Address Claim Procedure

While other higher layer protocols based on Controller Area Network (CAN) do not support node address assignments per default, the SAE J1939 protocol provides yet another ingeniously designed feature to uniquely identify Electronics Control Units (ECU) and their primary function.

Note: *As a reminder, a J1939 network node is usually called an "ECU." Each ECU can maintain more than one "Control Application (CA)." For that reason, the technical requirements usually refer to a CA rather than an ECU.*

2.4.2.1 Technical Requirements

The SAE J1939/81 Standard lists a number of technical requirements for the address claiming process:

1. Each control application (CA) must be capable of providing its unique 64-bit NAME.

2. CAs must successfully claim an address prior to sending messages (other than those for claiming an address) into the network.

3. The inability to successfully claim an address must be handled and reported to the network.

4. The CA must follow the network initialization associated with the address claiming process.

5. The CA must support a minimum set of network management requirements, including required responses to power interruptions.

2.4.2.2 Device NAME

The J1939 Standard defines a 64-Bit NAME, as shown in the following image, to uniquely identify an ECU in the network.

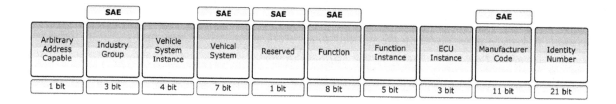

As indicated in the above image, some parameters and are assigned by the SAE, the rest is set by the manufacturer/developer of the application.

- **Arbitrary Address Capable** – Indicates whether or not the ECU/CA can negotiate an address (1 = yes; 0 = no). Some ECUs can only support one address; others support an address range.

- **Industry Group** – These codes are associated with particular industries such as on-highway equipment, agricultural equipment, and more.

- **Vehicle System Instance** – Assigns a number to each instance on the Vehicle System (in case you connect several networks – e.g. connecting cars on a train).

- **Vehicle System** – Vehicle systems are associated with the Industry Group and they can be, for instance, "tractor" in the "Common" industry or "trailer" in the "On-Highway" industry group.

- **Reserved** – Always zero.

- **Function** – This code, in a range between 128 and 255, is assigned according to the Industry Group. A value between 0 and 127 is not associated with any other parameter.

- **ECU Instance** – A J1939 network may accommodate several ECUs of the same kind (i.e. same functionality). The Instance code separates them.

- **Manufacturer Code** – The 11-Bit Manufacturer Code is assigned by the SAE.

- **Identity Number** – This field is assigned by the manufacturer, similar to a serial number, i.e. the code must be uniquely assigned to the unit.

For test and simulation purposes and with the assumption that the NAME is only used for the address claiming process, it is possible to set all required parameters to your liking. However, if you connect your application to a real-world SAE J1939 network, it is strongly advised to follow the official settings, which includes obtaining a manufacturer code.

For our Arduino projects, I have assigned the NAME fields in a way that they will not interfere when used within an existing vehicle network. This has been done by setting the *Identity Number* and *Manufacturer Code* to the maximum value, which will result in a more passive role during the address claim process. An ECU with a NAME of higher value is more likely to lose the competition with another node using the same address.

Note: *All settings as shown are used for demonstration purposes only. In all consequence, you must follow the SAE's recommendations. Also, you alone (and not the author or publisher) are responsible for the final implementation and the results thereof.*

```
#define NAME_IDENTITY_NUMBER                0xFFFFFF
#define NAME_MANUFACTURER_CODE              0xFFF
#define NAME_FUNCTION_INSTANCE              0
#define NAME_ECU_INSTANCE                   0x00
#define NAME_FUNCTION                       0xFF
#define NAME_RESERVED                       0
#define NAME_VEHICLE_SYSTEM                 0x7F
#define NAME_VEHICLE_SYSTEM_INSTANCE        0
#define NAME_INDUSTRY_GROUP                 0x00
#define NAME_ARBITRARY_ADDRESS_CAPABLE      0x01
```

Note: *These settings can be found in the ARD1939.h file of each project that involves the ARD1939 library.*

2.4.2.3 Preferred Address

For the purpose of a quick address claiming process, each control application should maintain a preferred address. SAE J1939/81 recommends that the preferred address (i.e. the address the ECU/CA attempts to claim on power-up) should be re-programmable to permit the proper configuration of the entire vehicle network. It also helps to prevent delays/problems during the address claiming procedure. The proper procedure would be to determine all preferred addresses in the entire network and design the addresses in a way that no collisions will occur.

According to the SAE J1939/81 Standard, each CA should attempt to use preferred addresses assigned by the SAE in correspondence to the *Industry Group*, meaning each Industry Group comes with a list of preferred addresses according to the CA's function.

In the case of the *Global* Industry Group, these addresses are in the range of 0 to 84, while the range of 128 to 247 is Industry Group specific. According to the SAE J1939/81 Standard the range of 85 through 127 are reserved for future assignment by the SAE.

Also, according to SAE J1939/81, a control application (CA) claiming a preferred address in the range 0 to 127 and 248 to 253 must perform the function defined for that process and specify that function within its NAME.

All these definitions may be confusing, even contradictory to the engineer when it comes to testing a J1939 ECU. For demonstration purposes (i.e. in our Arduino projects), we will assume a preferred address of 128 (0x80) and a negotiable address range between 129 and 247.

2.4.2.4 Address Claiming Procedure

In the following, we will focus merely on two basic messages, *Request Message* and *Address Claimed* (Yet again, for further, more detailed information see the literature recommendation in the appendix).

The *Request Message* is used by a CA to request information, such as NAME and addresses, from all other CAs in the entire network. Upon receipt of the *Request Message for Address Claimed*, each CA will transmit an *Address Claimed* message that contains the requested information. This procedure can be used to gather information from the network either for diagnosis/display services or even to determine a suitable node address.

Another method to set the node address is to send the *Address Claimed* message into the bus. This is the equivalent to interrogate whether or not the preferred address is already taken. If no response is received, it is safe to assume the address is available. If there is a response, the CA must modify the address and send another *Address Claimed* message.

The following flowchart demonstrates the address claim procedure:

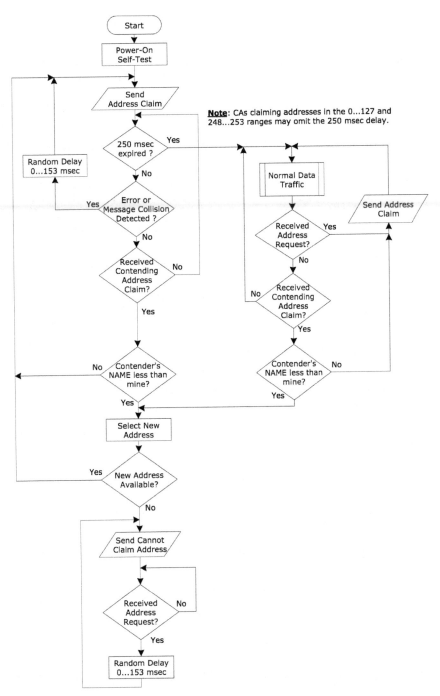

Note: The Address Claim procedure as shown in the image is fully implemented in the ARD1939 protocol stack library.

2.5 SAE J1939 Compliance Criteria

One question that has been raised repeatedly is the one about SAE J1939 compliance. Unfortunately, the SAE J1939 Standards Collection does not define at what point a device can truly be considered compliant, and there is no official test document or test facility.

There is an increasing number of SAE J1939 devices and software available in the market, but most of them fail to reveal their level of compliance with the SAE J1939 Standard. Yet, others confuse with a list of supported J1939 standards in an attempt to separate them from the competition.

For that reason let's have a look at the official SAE documents: The referenced standards (e.g. SAE J1939/21, SAE J1939/81, etc.) are part of the J1939 Standards Collection. The "SAE Truck and Bus Control & Communications Network Standards Manual – 2007 Edition" is a colossal work of roughly 1600 pages where about 1000 pages refer to data such as Parameter Group Assignments, Address and Identify Assignments, Parameter Group Numbers, and more.

It was designed to follow the ISO/OSI 7-Layer Reference Model, where each layer is addressed by a corresponding document as demonstrated in the following image.

7. Application Layer	J1939/7x
6. Presentation Layer	J1939/6x
5. Session Layer	J1939/5x
4. Transport Layer	J1939/4x
3. Network Layer	J1939/3x
2. Data Link Layer	J1939/2x
1. Physical Layer	J1939/1x

The SAE has named documents addressing the transport (4), session (5), and presentation (6) layer in the ISO/OSI 7-Layer Reference Model. These layers are, however, not documented (in all consequence they are not necessary for J1939) and thus the corresponding documents have not been created.

The SAE J1939/1x and SAE J1939/2x standards describe the physical layer (wiring) and the data link layer (CAN controller functions). In addition, the SAE J1939/21 standard describes the transport protocol (TP), which is responsible for the transport of messages with more than eight data bytes. The SAE J1939/81 standard describes the address claim process. In all truth (and I mentioned this earlier), that's all an SAE J1939 application needs.

In other words, any SAE J1939 hardware must support SAE J1939/1x and SAE J1939/21, otherwise they're useless, because these standards describe the CAN bus physical layer and the basic protocol features. Any SAE J1939 software/firmware must support SAE J1939/81 (address claim process) and, if needed, SAE J1939/21 (transport of up to 1785 bytes per message).

Note: *The implementation of the Transport Protocol (TP), i.e. the transport of up to 1785 data bytes in a message, is highly application-specific. Some ECUs and their control applications (CAs) will need it, some won't.*

Everything beyond SAE J1939/21 and SAE J1939/81 (e.g. SAE J1939/73 diagnostics, which belongs to the application layer) is cream on the cake but is not part of the actual protocol.

Some vendors in the J1939 business claim "compliance" with the SAE J1939/71 standard, which is a mere marketing hype. SAE J1939/71 describes the Parameter Group Numbers (PGNs), but while PGNs are being transmitted per J1939, they are, with a few exceptions, not features of the protocol. PGNs for protocol management such as transport protocol (TP) and address claim are defined in SAE J1939/21 and SAE J1939/81.

Caution is also advised in regards to an SAE J1939/81 (address claim procedure) compliance claim. Many devices in the marketplace will work only with one hard-coded node ID (which is still compliant with the standard). The problem starts, however, when that single address conflicts with an existing vehicle network where that particular address is already taken (I have worked with devices that didn't care and continued sending data, even though their address conflicted with that of another device).

Note: *The ARD1939 protocol stack, as introduced in a later chapter, supports the full SAE J1939/81 standard. It can negotiate a single address or an address range and removes itself from the bus in case of an address conflict. It also supports the full Transport Protocol (TP) according to SAE J1939/21.*

3. SAE J1939 Applications with the Arduino

In general, there are three different intentions for connecting to a J1939 vehicle network:

1. Mere monitoring, processing, and display of network data traffic.
2. All functions as described under 1. but extended by the ability of sending data into the J1939 bus.
3. All functions as described under 1. and 2. but extended by the J1939 Transport Protocol, supporting messages with more than 8 data bytes.

Let's address these three approaches:

1. The mere monitoring, processing, and display of network data traffic does <u>not</u> require any network management functionality (i.e. the actual SAE J1939 protocol). Simply connect your monitoring device to the vehicle bus and filter those messages relevant to your needs.
2. The ability of sending or requesting data to/from the J1939 bus requires that your system (ECU = Electronic Control Unit) owns an 8-bit address. Since addresses in a J1939 vehicle network are dynamic (i.e. they may change after each power-up), your system must follow the J1939 Address Claiming Procedure according to SAE J1939/81.
3. The J1939 Transport Protocol (SAE J1939/21) is that particular part of the network management functionality that raises the level of complexity. However, the vast majority of J1939 applications do not require the Transport Protocol (TP). The TP is mostly used for non-critical, data-intensive communication tasks.

The following chapter is about J1939 monitoring and simulation projects, which do not require a J1939 protocol implementation. A later chapter will introduce the ARD1939 protocol stack, allowing full communication with a J1939 vehicle network.

3.1 SAE J1939 Monitoring And Simulation

As I wrote before, both versions of the Arduino used in this book, the Uno and the Mega 2560, are well suitable for SAE J1939 monitoring and simulation projects. Until now, the simulation of J1939 data traffic involved the purchase of expensive simulation devices and the associated steep learning curves. All this changes with the Arduino's easy-to-use hardware and software solutions.

Writing J1939 monitoring and simulation examples for the Arduino is a matter of a few hours or even less when using the following programming samples as a template.

Note: *All Arduino projects (Sketches) as introduced in the following will use the Serial Monitor at a baud rate of 115,200 bit/sec (See also my remarks in chapter "Serial Interface And CAN Timing Considerations").*

3.1.1 SAE J1939 Simulation Example

The following programming sample demonstrates effectively how easy simulating SAE J1939 data traffic can be (while the level of complexity can be easily adjusted to the application requirements).

However, before we dive into the actual coding, we need to design the simulated messages, i.e. we need to determine the CAN 29-Bit message ID, which includes the actual PGN, the PGN's priority, and the application's source address.

3.1.1.1 Message Design

For this programming sample I have randomly chosen two PGNs from the vast selection of available parameter group numbers. They are:

- **PGN 65267** – Vehicle Position, provides the vehicle's latitude and longitude.
  - **Transmission Repetition Rate**: 5 sec
  - **Parameter Group Number**: 0xFEF3
  - **Data Length**: 8 bytes
  - **Default Priority**: 6

- **PGN 65269** – Ambient Conditions, provides barometric pressure, cab interior temperature, ambient air temperature, engine air inlet temperature, and road surface temperature.
  - **Transmission Repetition Rate**: 1 sec
  - **Parameter Group Number**: 0xFEF5
  - **Data Length**: 8 bytes
  - **Default Priority**: 6

Note: *In order to implement other or further PGNs, you will need a copy of the SAE J1939/71 standard for a detailed description of the parameters and their data format. You may find some of them by browsing the Internet, but be aware that there is no complete online reference. A description of all PGNs is out of the scope of this book.*

Besides the actual PGNs, we will also need to simulate the application's source address. The source address is usually determined through the address claim process of the SAE J1939 protocol stack, which is not yet part of this programming sample.

For mere test and demonstration purposes, it is sufficient to assume the addresses. In the following, we use 0x20 for PGN 65267 and 0x30 for PGN 65269.

Applying the PGNs plus priority plus source address, the message IDs are:

- **PGN 65267 – Vehicle Position: 0x18FEF320**

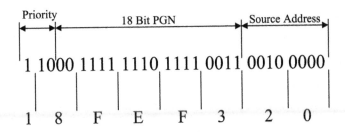

- **PGN 65269 – Ambient Conditions: 0x18FEF530**

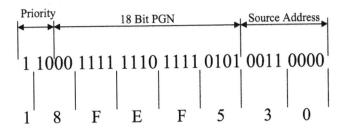

3.1.1.2 Arduino Code

The Arduino code is quite simple and virtually self-explanatory. However, let me lose a few words on the program's structure.

First of all, depending on the CAN shield used for this programming sample, please make sure you set the proper CS (Chip Select) for the CAN controller as described in module can.cpp (See also my remarks in the chapter about the CAN-BUS Shield by Seeed Studio).

In the main program module J1939_Data_Traffic_simulation I defined two messages, msgVehiclePosition and msgAmbientConditons, and filled them with random data.

In the main *loop* I use the *delay* function to create a one-second timer by means of the variable *nCounter*. The content of *nCounter* triggers the transmission of the corresponding messages.

The actual program code is fairly simple:

```
void loop()
{
  // Establish the timer base
  delay(1000);     // 1 sec
  nCounter++;

  // Send out PGN 65629 - Ambient Conditions
  canTransmit(0x18FEF530, msgAmbientConditions, 8);

  // Send out PGN 65627 - Vehicle Position
  if(nCounter == 5)
  {
    canTransmit(0x18FEF320, msgVehiclePosition, 8);
    nCounter = 0;  // Reset the counter

  }// end if

}// end loop
```

Note: *This Arduino project is available through the download page at http://ard1939.com.*

3.1.1.3 Proof of Concept

The result was tested with an ADFweb CAN-to-USB Gateway in combination with its Windows-based CAN analyzer tool. As the screen shot below shows, the PGN 65267 (Vehicle Position: 0x18FEF320) is received every 5 seconds, while PGN 65269 (Ambient Conditions: 0x18FEF530) appears every second (according to the time stamps).

NR	TIME	ID (HEX)	DATA (HEX)	ASCII
1	12:15:48 AM.268.1	18FEF530	38 37 36 35 34 33 32 31	87654321
2	12:15:49 AM.266.5	18FEF530	38 37 36 35 34 33 32 31	87654321
3	12:15:50 AM.264.9	18FEF530	38 37 36 35 34 33 32 31	87654321
4	12:15:51 AM.263.2	18FEF530	38 37 36 35 34 33 32 31	87654321
5	12:15:52 AM.261.6	18FEF530	38 37 36 35 34 33 32 31	87654321
6	12:15:52 AM.262.1	18FEF320	31 32 33 34 35 36 37 38	12345678
7	12:15:53 AM.260.2	18FEF530	38 37 36 35 34 33 32 31	87654321
8	12:15:54 AM.258.5	18FEF530	38 37 36 35 34 33 32 31	87654321
9	12:15:55 AM.256.9	18FEF530	38 37 36 35 34 33 32 31	87654321
10	12:15:56 AM.255.3	18FEF530	38 37 36 35 34 33 32 31	87654321
11	12:15:57 AM.253.6	18FEF530	38 37 36 35 34 33 32 31	87654321
12	12:15:57 AM.254.2	18FEF320	31 32 33 34 35 36 37 38	12345678
13	12:15:58 AM.252.2	18FEF530	38 37 36 35 34 33 32 31	87654321
14	12:15:59 AM.250.6	18FEF530	38 37 36 35 34 33 32 31	87654321

3.1.1.4 Advanced Program Version

While the previous programming sample proves the simplicity of developing an SAE J1939 data traffic simulator, I deemed it necessary to improve the code in way that makes simulation extensions and modifications easier. The two elements that needed improvement were the message transmission (which requires assembling the message ID from PGN, priority and source address) and the timer control (which is quite rudimentary).

In this advanced version I replaced the *canTransmit* function with *j1939Transmit*, which allows passing the PGN, priority, and source address as separate parameters.

So, instead of:

- `canTransmit(0x18FEF530, msgAmbientConditions, 8);`
- `canTransmit(0x18FEF320, msgVehiclePosition, 8);`

we use:

- `j1939Transmit(65269, 6, 0x30, msgAmbientConditions, 8);`
- `j1939Transmit(65267, 6, 0x20, msgVehiclePosition, 8);`

The timer control adds a higher level of complexity, but it will improve the adaptability of the code. First of all, we need to declare the timers to be used:

- `struct j1939Timer pTimerVehiclePosition;`

In the *setup* routine, the timers need to be initialized:

- ```
  j1939TimerReset(&pTimerVehiclePosition);
  pTimerVehiclePosition.nCount = 5000;
  pTimerVehiclePosition.bStart = true;
  ```

In the *loop* function, we must call the timer control function, which uses a time base of 1 millisecond:

- `j1939TimerControl();`

Also, in the loop function, we must check the timer:

- ```
  if(pTimerVehiclePosition.bExpired == true)
  {
      j1939TimerReset(&pTimerVehiclePosition); // Reset the timer
      j1939Transmit(65267, 6, 0x20, msgVehiclePosition, 8);  // Transmit
      pTimerVehiclePosition.nCount = 5000; // Restart the timer
      pTimerVehiclePosition.bStart = true;
  }// end if
  ```

51

The modified timer version provides a finer timer resolution, which also improves any reaction time of the application, for instance, when it comes to receiving SAE J1939 message frames and modifying the transmitted data according to the received information.

In this case, I refrain from showing another screen shot as a proof of concept. The image is identical to the one in the previous chapter.

Note: *This Arduino project is available through the download page at http://ard1939.com.*

3.1.1.5 SAE J1939 Stress Simulator

The increased flexibility in combination with the finer timer resolution we reached in the previous project allows us the easy implementation of another task, namely a stress simulator.

Stress simulator, in this case, means increasing the vehicle bus load and thus "stress testing" a certain J1939 device. The question for many developers is whether or not their device can handle increased busload and still function according to specification.

As a fair warning upfront, the following project has its limits in terms of reaching maximum busload (100%) due to hardware limitations (Arduino clock time) and the maximum timer resolution of 1 millisecond. The average J1939 message takes roughly 540 microseconds to transmit, and sending it every one millisecond will result in a busload of maximal 50%.

Sending two consecutive messages within one millisecond cycle can easily increase the busload, but this is also the case where we reach the hardware limitations. Further improvements could be accomplished through implementing an interrupt service routine for message transmission and a timer resolution finer than that provided by the *delay()* function. This would, however, result in a great deal of code tweaking, and the efforts would not be worth the result.

The easiest solution to maximum busload is actually running the same program on two Arduinos connected to the same bus.

Note: *An SAE J1939 busload of 60% is considered extremely high (100% would be continuous data flow) and is thus sufficient for a stress test.*

In the *J1939_Stress_Test* project, I have simply copied the previous project and use the same PGNs to transmit. However, in the *loop()* function I have decreased the timer values to 1 millisecond, meaning the program will send two messages within one millisecond. According to my ADFweb CAN analyzer software, this creates a busload of 57%.

In another variant, I loaded the same program onto another Arduino simultaneously connected to the same network, however, played with the number of transmitted PGNs and their frequency:

- 1 message every 2 milliseconds: Busload = 76%

- 1 message every 1 millisecond: Busload = 79%

- 2 messages every 1 millisecond: Busload = 79%

My assumption is that the ADFweb interface's limit is at 79%, which is not a bad value. Due to internal latency times, you will hardly find any CAN gateway that comes close to 100%, unless you are willing to spend some major money.

I will refrain from posting any of the project's code here, since it is almost identical to the previous one.

Note: *This Arduino project is available through the download page at http://ard1939.com.*

3.1.2 SAE J1939 Monitoring Examples

The following chapter is primarily about J1939 data monitoring, i.e. the Arduino projects as described work mainly in "Listen-Only" mode. Any bidirectional communication with the vehicle bus requires a J1939 protocol stack implementation, specifically the address claim process. Without a node (source) address, accessing the vehicle bus may cause address conflicts and malfunctions.

The following projects were designed with the assumption that using a presumed node address will not affect the network, either due to lab conditions or the knowledge of all node addresses in the vehicle network (thus preventing any address collision). Keep in mind that these projects are meant for mere demonstration and educational purposes.

3.1.2.1 Receiving J1939 Message Frames

This Arduino project provided a bit of a challenge after I had forgotten my own advice.

Note: *In order to test a CAN/J1939 application, you need at least two CAN/J1939 nodes to establish a network communication. The second node can be another Arduino with CAN shield or (if the budget allows) another CAN/J1939 device with CAN/J1939 data monitoring capabilities.*

In other words, in order to receive J1939 messages, I needed a transmitter. However, the solution was simple: I used a second Arduino with CAN shield and ran the application as described in chapter *SAE J1939 Simulation Example*, in this case the advanced version.

In order to receive and transmit J1939 message frames in a more comprehensive way (rather than receiving mere CAN messages that involve decoding the message ID), I added more J1939 functionality to the *can.cpp* module.

The new (or modified) function calls are:

- *j1939Transmit* – This function call (which we already used in the previous project) now supports the setting of source and destination address – the previous version allowed only the source address and thus did not directly support peer-to-peer communication.

- *J1939Receive* – This function's parameters are very similar to those as used with *j1939Transmit* with the difference that the parameters are pointers to variables, because these are parameters we receive.

- *canReceive* – This function (called by *j1939Receive*) has been extended to use a CAN message ring buffer in order to prevent timing problems and, consequently, avoid the loss of messages.

- *J1939PeerToPeer* – This function (called by both, j1939Transmit and j1939Receive) determines whether or not the passed PGN represents a peer-to-peer (direct node) communication. If yes, the function will insert the destination address into the CAN message ID.

The actual *loop()* function is fairly simple, since most of the code is used to display the received J1939 data frame.

```
// ------------------------------------------------------------------
// Main Loop - Arduino Entry Point
// ------------------------------------------------------------------
void loop()
{
  // Declarations
  byte nPriority;
  byte nSrcAddr;
  byte nDestAddr;
  byte nData[8];
  int nDataLen;
  long lPGN;

  char sString[80];

  // Check for received J1939 messages
  if(j1939Receive(&lPGN, &nPriority, &nSrcAddr, &nDestAddr, nData,
                  &nDataLen) == 0)
  {
        sprintf(sString, "PGN: 0x%X Src: 0x%X Dest: 0x%X ", (int)lPGN,
                nSrcAddr, nDestAddr);
        Serial.print(sString);
```

```
    if(nDataLen == 0 )
      Serial.print("No Data.\n\r");
    else
    {
      Serial.print("Data: ");
      for(int nIndex = 0; nIndex < nDataLen; nIndex++)
      {
        sprintf(sString, "0x%X ", nData[nIndex]);
        Serial.print(sString);

      }// end for
      Serial.print("\n\r");

    }// end else

  }// end if

}// end loop
```

Note: *This Arduino project is available through the download page at http://ard1939.com.*

The data is being displayed on the Arduino's Serial Monitor, and the result is as expected (for demonstration purposes, I have highlighted PGN 0xFEF3, which appears every five seconds).

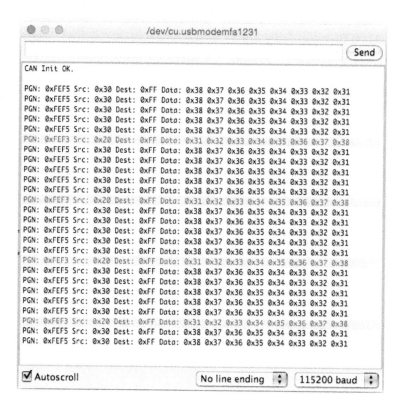

Note: *The destination address is being displayed as 0xFF (255). 255 is the global address, meaning that both PGNs are broadcast messages.*

3.1.2.2 Receiving and Responding to J1939 Request Frames

In this following project, we will discover yet another SAE J1939 protocol feature, namely the *Request Message* (as defined in the SAE J1939/21 standard). As with the previous project, we will need two J1939 nodes, and therefore will need two projects, one to receive the request and response, the second to send the request and receive the response.

But first, let's have a look at the *Request Message*: This message type, which is represented by a dedicated PGN, provides the means to request information globally (broadcast) or from a specific node address (peer-to-peer).

- **PGN 59904** – Request for Address Claimed Message.
    - o **Parameter Group Number**: 0xEAxx (where xx represents the destination address, either the global address 255 for broadcasting, or the specific node address)
    - o **Data Length**: 3 bytes
    - o **Data**: Bytes 1, 2, 3 = PGN being requested (LSB first, MSB last)
    - o **Default Priority**: 6

Note: *The transmission/reception of a 3-byte PGN according to the SAE J1939 standard always confuses programmers that are new to J1939, specifically the sending of LSB first and MSB last, which defies common programmer's thinking. To make things worse, there is only one small sentence within the massive 1600-page SAE J1939 Standards Collection that refers to that little, nevertheless important detail.*

A node receiving the *Request Message* will respond with the requested PGN and its data.

For our current project we will create the following scenario: Node 0x20 (32) sends out a request for PGN 65262 (0xFFFE – Engine Temperature) to node 0x30 (48). Node 0x30 will receive the request and responds by sending the PGN.

Note: *The node numbers and the requested PGN have been randomly chosen and serve only as demonstration examples. In all consequence, the method of requesting messages as shown here is not 100% J1939 compliant. One of the following projects, the J1939 Network Scanner, is a better example for the use of the Request Message.*

To re-iterate the tasks at hand:

1. Node 0x30 receives the request message and responds by sending the PGN
2. Node 0x20 sends the request message and receives the response

To simplify the project management, I have combined the two tasks into one project and loaded it onto two separate Arduino systems. In this case both system simulate both node addresses.

The following code, an excerpt from the *loop()* function, explains the function of this project:

```
// Check for received J1939 messages
if(j1939Receive(&lPGN, &nPriority, &nSrcAddr, &nDestAddr, nData,
                &nDataLen) == 0)
{
  // Display the received message on the serial monitor
  :
  :
  // Analyze the received PGN
  switch(lPGN)
  {
    case PGN_RequestMessage:

      if(nData[0] == PGN_EngineTemperatureLSB
      && nData[1] == PGN_EngineTemperature2ND
      && nData[2] == PGN_EngineTemperatureMSB)
      {
        // Request message for engine temperature was received
        j1939Transmit(PGN_EngineTemperature, 6, NODE_Response, NODE_Request,
                      msgEngineTemperature, 8);  // Transmit the message
      }

      break;

  }// end switch

}// end if

// Check for user input per serial monitor
// Send request message with user input
if(ReadSerialMonitorString() == true)
  j1939Transmit(PGN_RequestMessage, 6, NODE_Request, NODE_Response,
                msgRequest, 3);
  :
  :
```

1. As soon as we receive a J1939 message frame we display it on the serial monitor (code has been blended out).

2. We analyze the received PGN.

3. If the PGN is the *Request Message* PGN, we check the attached data for the requested PGN.

4. If the requested PGN is for Engine Temperature we transmit the corresponding PGN.

5. If the user sends any data through the serial monitor, we send out a request for Engine Temperature.

Note: *This Arduino project is available through the download page at http://ard1939.com.*

The following image shows the Arduino's serial monitor after the user has sent a request message. The other Arduino node answered the request.

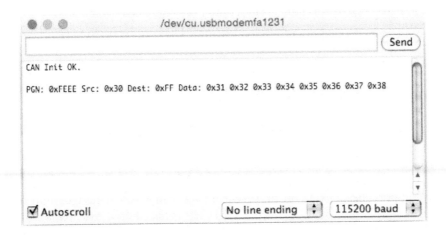

Note that the destination address is shown as 0xFF (255), which is the global address. As I mentioned previously, this example is not quite J1939 compatible. The destination address must be the global address, because the PGN 65262 (0xFEEE) cannot be transmitted from peer-to-peer but only as a broadcast message.

The following screen shot taken through the ADFweb CAN analyzer shows the data traffic on the J1939 vehicle bus, and it confirms the functionality of the project.

NR	TIME	ID (HEX)	DATA (HEX)	ASCII
1	12:52:11 AM.672.0	18EA3020	EE FE 00	îþ
2	12:52:11 AM.674.0	18FEEE30	31 32 33 34 35 36 37 38	12345678

Line 1
- 0x18 indicates a message priority of 6.
- 0xEA30 indicates a request message sent to node address 0x30.
- 0x20 represents the address of the requesting node (source address).

Line 2:

- 0x18 represents a message priority of 6.

- 0xFEEE is the PGN for Engine Temperature (as requested).

- 0x30 represents the address of the transmitting node.

3.1.3 A Simple SAE J1939 to USB Gateway

I deem it necessary to provide a definition of "SAE J1939 to USB Gateway," because there are several variants of gateway applications.

If you browse through the Internet in search for J1939 gateways, you will primarily find CAN gateways that support the conversion of 11-Bit and 29-Bit CAN messages into another serial technology such as USB, RS232, etc. Since SAE J1939 is based on CAN messages with a 29-Bit identifier, many manufacturers take the liberty of calling their devices "SAE J1939 gateways," which is not only misleading but also may lead to technical problems when connecting these devices to a real vehicle (diesel engine).

The problem lies in the fact that these so-called gateways do not come with an integrated SAE J1939 protocol stack and thus cannot support the address claiming process according to SAE J1939/81. When transmitting data into a J1939 vehicle network you need to supply a node (source) address with the data frame, and that node address must be unique. The absence of a protocol stack in the gateway may result in node ID collisions and therefore may cause serious technical problems.

The problem can be circumvented when all node IDs in the network are hard-coded and you assign a free node ID to messages you send into the network. This may work in limited cases, but this method is, naturally, not recommended.

The purpose of these types of J1939 gateways should be mere data monitoring, i.e. only receiving J1939 messages (you still need to decipher J1939 messages longer than 8 bytes, which is a cumbersome task). In order to communicate with the vehicle bus (bidirectional communication) you need a full SAE J1939 protocol implementation with address claim procedure and support of messages longer than 8 bytes.

Nevertheless, in the following I will introduce this kind of gateway with all its disadvantages, but I will do so for mere demonstration and educational purposes. The Arduino sketch, as introduced in the following, is the first step toward a full SAE J1939 gateway with integrated protocol stack (See chapter *ARD1939 – SAE J1939 Protocol Stack for Arduino*).

Some of the previous Arduino sketches as introduced in this book already provide the features for a simple SAE J1939 to USB gateway. However, they were designed to merely receive messages. A gateway application requires that we can send messages as well. The need for bidirectional communication also leads to the next requirement for a J1939 gateway, namely the definition of a communication protocol between the gateway and the host system (usually a PC).

Since our focus is on the Arduino hardware, the connection between the gateway (i.e. the Arduino) and the PC is established per USB interface. The foundation of the USB protocol is already established through the Serial.print function and the protocol will be based on transmitting ASCII text.

The data communication is managed through the Arduino's serial monitor, but I will also introduce a replica of the serial monitor with Visual Studio C# in another chapter to follow. If you wish to create a more sophisticated data communication between the Arduino and a PC, the following projects will provide a solid foundation.

Our simple J1939 to USB gateway will provide three functions:

- A J1939 Network Scanner, i.e. the Arduino will scan a J1939 network for existing nodes and retrieve information from them.

- A Data Traffic Monitor similar to the previously introduced projects.

- A Data Traffic Simulator, i.e. the Arduino will send user-defined messages into the network.

Note: *As usual, the following Arduino projects are available through the download page at http://ard1939.com. However, all three functions (network scanner, data traffic monitor, and data traffic simulator) are combined into one project, since the ladder two features are additions to the previous project.*

3.1.3.1 J1939 Network Scanner

Just like the previous project (See chapter *Receiving and Responding to J1939 Request Frames*) we will be using the SAE J1939 *Request Message*, in this case to inquire node addresses from the vehicle network. The Request Message (a.k.a. the *Request for Address Claimed Message*) is also used during the address claim process as defined in SAE J1939/81. Our network scanner project is very similar to the address claim process, since we are inquiring addresses from the network (The address claim process will inquire specific addresses to see if they are taken or not).

Note: *Yet again, this sample project will only work with two nodes, one Arduino loaded with the following project plus a second node loaded with the full ARD1939 – SAE J1939 Protocol Stack for Arduino (See also chapter ARD1939 – Implementation).*

Yet again, let's have a look at the *Request for Address Claimed Message*:

- **PGN 59904** – Request for Address Claimed Message
  - **Parameter Group Number**: 0xEAxx (where xx represents the destination address, either the global address 255 for broadcasting, or the specific node address)
  - **Data Length**: 3 bytes
  - **Data**: Bytes 1, 2, 3 = Requested PGN (LSB first, MSB last)
  - **Default Priority**: 6

The *Request for Address Claimed* message (when using a full protocol stack with its address claim procedure) is used to request the sending of an *Address Claimed* message from either a particular node in the network or from all nodes (use of global destination address = 255). The *Address Claimed* message (as described in the following) will provide the requested information, i.e. address and NAME of the responding node(s).

The purpose of sending such a request may be for several reasons, for instance:

- A node is checking whether or not it can claim a certain address.

- A node is checking for the existence of another node (Controller Application) with a certain function.

The response to a *Request for Address Claimed* message can be multiple:

- Any addressed node that has already claimed an address will respond with an *Address Claimed* message.

- Any addressed node that was unable to claim an address will respond with a *Cannot Claim Address* message.

- Any addressed node that has not yet claimed an address should do so by responding with their own *Address Claimed* message where the source address is set to NULL (254).

- A node sending the Request for Address Claimed message should respond to its own request in case the global destination address (255) was used.

The response to the *Request for Address Claimed* message is the actual *Address Claimed* message as defined in the following:

- **PGN 60928** – Address Claimed Message
  - **Parameter Group Number**: 0xEExx (where xx represents the destination address, either the global address 255 for broadcasting, or the specific node address)
  - **Data Length**: 8 bytes
  - **Data**: NAME of the control application (See also chapter SAE J1939/81 - Address Claim Procedure
  - **Default Priority**: 6

The code for our current Arduino project, the Network Scanner, is, yet again, fairly simple:

```
void loop()
{
  // Declarations
  byte nPriority = 0;
  long lPGN = 0;
  byte nSrcAddr = 0;
  byte nDestAddr = 0;
  byte nData[8];
  int nDataLen;

  // Send Request for Address Claimed with keystroke
  if(Serial.available() > 0)
  {
    Serial.setTimeout(1);
    Serial.readBytes(sString, 1);
    Serial.print("Network Scan Initiated:\n\r");
    nData[0] = PGN_AddressClaimedLSB;
    nData[1] = PGN_AddressClaimed2ND;
    nData[2] = PGN_AddressClaimedMSB;
    j1939Transmit(PGN_RequestMessage,
                  PRIORITY_RequestMessage, SA,
                  GLOBALADDRESS, nData, 3);
  }

  // Check for received J1939 messages
  if(j1939Receive(&lPGN, &nPriority, &nSrcAddr, &nDestAddr, nData,
                  &nDataLen) == 0)
  {
    switch(lPGN)
    {
      case PGN_AddressClaimed:
        sprintf(sString, "Addr Claimed: 0x%X %dd\n\r", nSrcAddr, nSrcAddr);
        Serial.print(sString);
        break;

      // We'll add more code here for the next project...

    }// end switch
```

```
}// end if

}// end loop
```

- As soon as the user enters an input through the serial monitor, we transmit the *Request for Address Claimed* message. We use the global address (255) to receive responses from all nodes in the network.

- We check the CAN interface for a received J1939 data frame.

- If a message was received, we check if it was an *Address Claimed* message.

- If the message was the *Address Claimed* message, we print the address as received.

My configuration allowed me to connect two SAE J1939 nodes to the network, an ARM system at address 0x2b (43) and an Arduino Mega 2560 at address 0x80 (128).

The following screen shot from the aerial monitor confirms the configuration and the addresses:

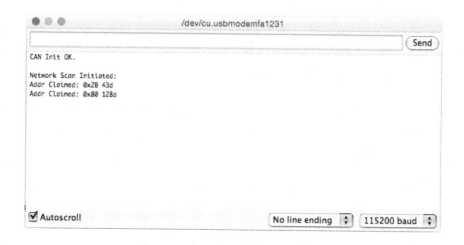

As another proof of concept, let's have another look at the ADFweb analyzer software, which I have always connected:

NR	TIME	ID (HEX)	DATA (HEX)
1	12:11:56 AM.134.2	18EAFF33	00 EE 00
2	12:11:56 AM.134.8	18EEFF2B	00 00 60 43 00 FF FE 90
3	12:11:56 AM.135.5	18EEFF80	00 00 E0 FF 00 FF FE 10

- **Line 1:** Our Arduino application sends out the *Request for Address Claimed* message. Note the ID where 18 indicates a priority of 6, EAFF represents the *Request for Address Claimed* message including the global address 255 (0xFF). The last byte of 0x33 represents the simulated source address of our application. The data field contains the three-byte PGN for *Address Claimed* message (00E000; LSB first, MSB last).

- **Line 2:** The node at address 0x2B (43) responds with the *Address Claimed* message. Note the ID where 18 indicates a priority of 6, EEFF represents the *Address Claimed* message including the global address 255 (0xFF). The last byte represents the node's source address. The data field contains the eight-byte NAME of the control application (ECU).

- **Line 3:** Same as line 2 but different address and NAME.

In a next step, to demonstrate the difference between message broadcasting and peer-to-peer, I replaced GLOBALADDRESS in the *j1939Transmit* function with 0x80, which means that we are checking the network for the existence of one particular address.

The result on the Arduino serial monitor is not very impressive after all; it shows (as expected) only one response:

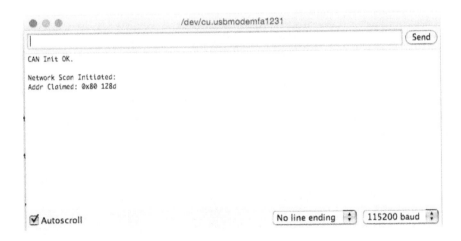

The ADFweb CAN analyzer software, however, reveals more detail about the peer-to-peer (node to node) communication:

NR	TIME	ID (HEX)	DATA (HEX)
1	12:37:08 AM.949.0	18EA8033	00 EE 00
2	12:37:08 AM.950.1	18EE3380	00 00 E0 FF 00 FF FE 10

- **Line 1:** Our Arduino application sends out the *Request for Address Claimed* message. Note the ID where 18 indicates a priority of 6, EAFF represents the *Request for Address Claimed* message including the inquired address 0x80 (128). The last byte of 0x33 represents the simulated source address of our application. The data field contains the three-byte PGN for *Address Claimed* message (00E000, LSB first, MSB last).

- **Line 2:** The node at address 0x80 (128) responds with the *Address Claimed* message. Note the ID where 18 indicates a priority of 6, EE33 represents the Address Claimed message including the requestor address 0x33. The last byte represents the node's source address. The data field contains the eight-byte NAME of the control application (ECU).

3.1.3.2 J1939 Data Traffic Monitoring

In the following Arduino sketch, "data traffic monitoring" means nothing else but receiving and displaying J1939 data messages (Refer also to chapter *Receiving J1939 Message Frames*).

Rather than creating a new project, we will only need to add a few code lines to the previous network scanner project, specifically the *switch(lPGN)* section (refer to the highlighted code):

```
switch(lPGN)
{
  case PGN_AddressClaimed:
    sprintf(sString, "Addr Claimed: 0x%X %dd NAME: ", nSrcAddr, nSrcAddr);
    Serial.print(sString);
    for(int nIndex = 0; nIndex < nDataLen; nIndex++)
    {
      sprintf(sString, "0x%X ", nData[nIndex]);
      Serial.print(sString);
    }
    Serial.print("\n\r");
    break;

  default:
    sprintf(sString, "PGN: 0x%X Src: 0x%X Dest: 0x%X ", (int)lPGN,
            nSrcAddr, nDestAddr);
    Serial.print(sString);
    if(nDataLen == 0 )
```

```
      Serial.print("No Data.\n\r");
   else
   {
     Serial.print("Data: ");
     for(int nIndex = 0; nIndex < nDataLen; nIndex++)
     {
       sprintf(sString, "0x%X ", nData[nIndex]);
       Serial.print(sString);
     }
     Serial.print("\n\r");

   }// end else

   break;

}// end switch
```

After filtering the Address Claimed message, we also allow any other PGN we receive from the vehicle bus. In this case, we display not only the PGN but also the source address (transmitting node), the destination address (receiving node), and the actual data.

Note: *For variations of this project, you can filter more, specific PGNs or even filter messages by, for instance, destination address (where destination address is the address of this control application).*

In order to test the sketch, we will, yet again, need another J1939 node, i.e. another Arduino with CAN shield to create the data traffic. In the following example, I used the project as described in chapter *SAE J1939 Simulation Example*. The result is just as expected:

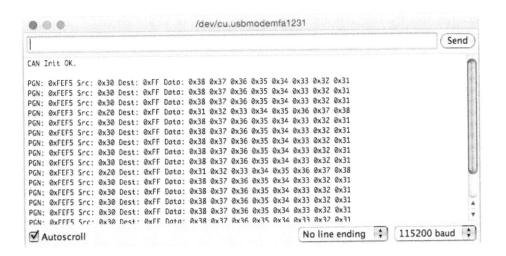

3.1.3.3 J1939 Data Traffic Simulation

In a previous project (see chapter *SAE J1939 Simulation Example*) I demonstrated the simulation of SAE J1939 data traffic. However, the messages in that example were transmitted within a certain frequency. In the following, we will transmit messages according to user input.

As before, we will use the previous sketch and add code without removing the previously added functionality. But, since we will transmit data according to user input, we need to separate the network scanner from the data transmission (Remember: the network scan was initiated by <u>any</u> user input).

From the previous chapter's sketch, I removed (actually, I commented out) the following code:

```
if(Serial.available() > 0)
{
  Serial.setTimeout(1);
  Serial.readBytes(sString, 1);
  Serial.print("Network Scan Initiated:\n\r");
  nData[0] = PGN_AddressClaimedLSB;
  nData[1] = PGN_AddressClaimed2ND;
  nData[2] = PGN_AddressClaimedMSB;
  j1939Transmit(PGN_RequestMessage, PRIORITY_RequestMessage, SA,
                GLOBALADDRESS, nData, 3);

}// end if
```

The new sketch now contains a function call *ReadSerialMonitorString* that returns the text string the user entered plus the length of the entry. For the network scan I chose "s" as the command. Consequently, the code looks like:

```
// Check for user input per serial monitor
int nCount = ReadSerialMonitorString(sString);
if(nCount > 0)
{
  if(nCount == 1 && sString[0] == 's')
  {
    Serial.print("Network Scan Initiated:\n\r");
    nData[0] = PGN_AddressClaimedLSB;
    nData[1] = PGN_AddressClaimed2ND;
    nData[2] = PGN_AddressClaimedMSB;
    j1939Transmit(PGN_RequestMessage, PRIORITY_RequestMessage, SA,
                  GLOBALADDRESS, nData, 3);

  }// end if
  :
  : More code to follow here...
  :
}// end if
```

The following screen shot shows the same result as in chapter *J1939 Network Scanner*, but, for the sake of adding some more information, I have added the code to print the control application's NAME.

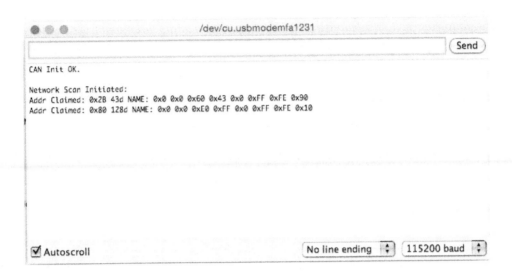

As I mentioned previously, in order to accomplish basic gateway functionality with the Arduino, we will need to establish a communications protocol between the Arduino's USB interface and the PC. The foundation of this USB protocol is already established through the Serial.print function and the protocol will be based on transmitting ASCII text.

In addition, we need to specify the commands to send out J1939 data according to user input. As described before, we use the letter "s" to initiate a network scan.

For the sending of a full SAE J1939 data frame, I have defined the following user input format based on text (ASCII) input:

- **PGN** – 4 characters
- **Priority** – 1 character
- **Source Address** – 2 characters
- **Destination Address** – 2 characters
- **Data** – 2 to 16 characters (representing 1 up to 8 data bytes)

All parameters will be separated by a space character, and the assumption is that all parameters are entered in hex format.

Example: EAFF 6 33 FF 00EE00

69

Per the previous example (*Request for Address Claimed* message), we are sending PGN 0xEAFF with a priority of 6, a source address of 0x33, the global destination address (255), and three data bytes.

Note: *The following code accomplishes only a very rudimentary user input verification. Please feel free to refine the code or add more functionality to the protocol. Additional features could be, for instance, the adding/deleting of PGN filters (allowing only user-defined PGNs).*

The code as shown is an extension to the previous code excerpt (see highlighted section).

```
int nCount = ReadSerialMonitorString(sString);
if(nCount > 0)
{
  if(nCount == 1 && sString[0] == 's')
  {
    Serial.print("Network Scan Initiated:\n\r");
    nData[0] = PGN_AddressClaimedLSB;
    nData[1] = PGN_AddressClaimed2ND;
    nData[2] = PGN_AddressClaimedMSB;
    j1939Transmit(PGN_RequestMessage, PRIORITY_RequestMessage, SA,
                  GLOBALADDRESS, nData, 3);

  }// end if
  else if(ParseUserCommand(sString, nCount, &lPGN, &nPriority, &nSrcAddr,
          &nDestAddr, nData, &nDataLen) == true)
  {
    // Print some feedback for verification purposes
    sprintf(sString, "Transmitting PGN: 0x%X P: 0x%X SA: 0x%X, DA: 0x%X Data:
                     ", (int)lPGN, nPriority, nSrcAddr, nDestAddr);
    Serial.print(sString);
    for(int nIndex = 0; nIndex < nDataLen; nIndex++)
    {
      sprintf(sString, "0x%X ", nData[nIndex]);
      Serial.print(sString);
    }// end for
    Serial.print("\n\r\n\r");

    // Transmit the J1939 message
    j1939Transmit(lPGN, nPriority, nSrcAddr, nDestAddr, nData, nDataLen);

  }// end else if
  else
  {
    // Incorrect user input; print corresponding message
    Serial.print("Incorrect Data Input:");
    Serial.print(sString);
    Serial.print("\n\r\n\r");

  }// end else

}// end if
```

You may notice a new function called *ParseUserCommand*, which analyzes the user input and returns the result (true = correct, false = incorrect). When the input was found correct, the function returns the PGN, Priority, Source Address (SA), Destination Address (DA), and the actual data.

As a feedback and reference to the user, the J1939 message is printed on the serial monitor and then being transmitted into the J1939 network.

The following screen shot shows the serial monitor, and, as a proof of concept, I had entered two messages.

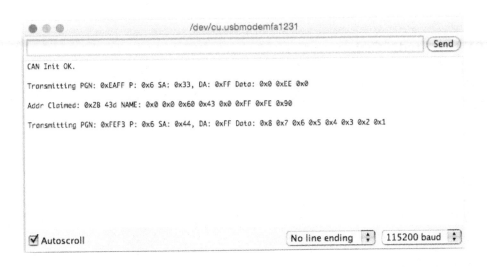

The first message is the *Request for Address Claimed* message, and therefore the serial monitor shows the reaction from the network (Node address 0x2B). The second message is the *Vehicle Position* message (PGN 65267 = 0xFEF3), which we had used in a previous example (See chapter SAE J1939 Simulation Example).

The following screen shot from the ADFweb CAN analyzer software confirms the functionality:

NR	TIME	ID (HEX)	DATA (HEX)
1	12:42:53 AM.894.5	17EAFF33	00 EE 00
2	12:42:53 AM.895.1	18EEFF2B	00 00 60 43 00 FF FE 90
3	12:46:18 AM.771.4	17FEF344	08 07 06 05 04 03 02 01

- **Line 1**: The Arduino sends the *Request for Address Claimed* message.
- **Line 2**: The SAE J1939 node at address 0x2B responds with the *Address Claimed* message.
- **Line 3**: The Arduino sends the *Vehicle Position* message (PGN 65267).

As I mentioned before, while this project is only a simple example, there are numerous possibilities to extend the code to a professional SAE J1939 solution. However, one important part of an SAE J1939 to USB gateway is the visualization of received and transmitted data. The following code will give you the basis for accomplishing a professional graphical user interface under Windows.

3.1.3.4 Visual Studio Code

While programming the Arduino can be exciting (especially since everything works so smoothly), the real fun comes when you can extend the Arduino's reach to a PC running Windows.

Note: *My apologies to all Mac and LINUX users for bringing a Windows programming example, but there is no better programming than using C# under Microsoft's Visual Studio. I have enjoyed programming under OS-X and LINUX, but when it comes to producing quick and effective programming examples, I prefer to stay with Visual Studio. However, the experienced programmer should be able to replicate the functionality of the serial monitor.*

To learn more about serial port programming (RS-232 and USB) under LINUX see http://www.teuniz.net/RS-232/. I consider this by far the most professional application for serial ports under LINUX. It also suits Windows applications but is primarily meant for compilers inferior to Visual Studio and/or for programming embedded systems.

In the following we assume that you have the Arduino USB driver installed under your Windows machine. The driver is automatically installed with the Arduino development environment.

As I have mentioned in my note, I am using Microsoft's Visual Studio 2012, and I have designed the following GUI that may look very familiar to the Arduino developer. Basically, this very simple program is a replica of the Arduino serial monitor.

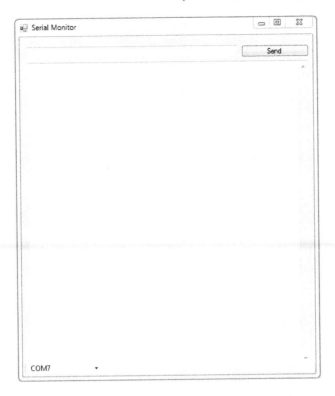

Note: *The following programming example will enable you to create your own graphical user interface. It demonstrates all the basics, such as how to read data from the PC's USB port, write data, and display data on the screen.*

The screen elements are a textbox for data entry, a command button to send the entry to the Arduino, and another larger text box to display the data coming from the Arduino. Last, but not least, there is a combo box displaying all available USB COM ports (It is your task to determine the proper USB port; there is no auto detection).

What the program does not provide is the baud rate settings, which has been hard-coded as 115,200 baud into the program but can be modified easily. Of course, this is not necessarily the professional way of doing it, but, after all, this programming sample serves as an example on reading/sending messages from/to the Arduino.

In regards to a more professional network monitoring and diagnostics tool, you will need more and different screen elements, and the baud rate settings should be part of that project. So feel free adding more functionality as you see fit.

All screen elements in this project stick with their default settings, however with a few exceptions as shown in the following:

Element	Name	Modified Property	Events
Form	Form1	Text = "Serial Monitor"	-
Text	txtSend	-	-
Text	txtReceived	Multiline = True Scrollbars = Vertical	-
Button	btnSend	Text="Send"	Click
ComboBox	cboCOMPort	-	SelectedIndexChanged

The following shows the C# program listing (the entire program is within the form):

```csharp
using System;
using System.Collections.Generic;
using System.ComponentModel;
using System.Data;
using System.Drawing;
using System.Linq;
using System.Text;
using System.Threading.Tasks;
using System.Windows.Forms;
using System.IO;
using System.IO.Ports;
using System.Threading;

namespace USBAccess
{
    public partial class Form1 : Form
    {
        // Constants
        public const int REC_BUFFER_SIZE = 500;
        public const int READ_TIMEOUT = 500;
        public const int WRITE_TIMEOUT = 500;
        public const int REC_BUFFER_FILLTIME = 80;

        public static SerialPort _serialport;

        public Form1()
        {
            InitializeComponent();

            string[] sPorts = new string[20];
            sPorts = SerialPort.GetPortNames();

            for (int nIndex = 0; nIndex < sPorts.Length; nIndex++)
                cboCOMPort.Items.Add(sPorts[nIndex]);

        }// end Form1

        // SetTextDeleg
        // -------------
        private delegate void SetTextDeleg(string text);

        // sp_DataReceived
        // ---------------
```

```
void sp_DataReceived(object sender, SerialDataReceivedEventArgs e)
{
    // Set the receive buffer size
    char[] sRecData = new char[REC_BUFFER_SIZE + 1];


    // Give the hardware some time to receive the whole message
    Thread.Sleep(REC_BUFFER_FILLTIME);

    try
    {
        int nBytes = _serialport.BytesToRead;

        // Read the string
        int nIndex;

        for (nIndex = 0; nIndex < nBytes; nIndex++)
        {
            int nRec = _serialport.ReadByte();
            sRecData[nIndex] = (char)nRec;

        }// end for

        sRecData[nIndex] = (char)0; // Terminate the string
        string sStr = new string(sRecData);

        // In case of RS232, this line causes a timeout,
        // meaning no data is being received
        this.BeginInvoke(new SetTextDeleg(si_DataReceived),
                                    new object[] { sStr });
    }
    catch (TimeoutException) { }

}// end _serialport_DataReceived

// si_DataReceived
// ---------------
private void si_DataReceived(string data)
{
    if(txtReceived.TextLength == 0)
        txtReceived.Text = data;
    else
        txtReceived.Text += "\n\r" + data;

    // Set cursor to end of screen
    txtReceived.SelectionStart = txtReceived.TextLength;
    txtReceived.ScrollToCaret();
    txtReceived.Refresh();

}// end si_DataReceived

// btnSend_Click
// -------------
private void btnSend_Click(object sender, EventArgs e)
{
    // Make sure the serial port is open before trying to write
```

75

```csharp
        try
        {
            if (!(_serialport.IsOpen))
                _serialport.Open();

            if (txtSend.Text.Length > 0)
                _serialport.Write(txtSend.Text);
            else
                MessageBox.Show("Please enter a message to be sent.",
                                                "Attention!");
        }
        catch (Exception ex)
        {
            MessageBox.Show("Error opening/writing to serial port." +
                                        ex.Message, "Error!");
        }

    }// end btnSend_Click

    // Event : cboCOMPort_SelectedIndexChanged
    //-----------------------------------------
    private void cboCOMPort_SelectedIndexChanged(object sender,
                                                EventArgs e)
    {
        // Define the serial port for the USB device
        _serialport = new SerialPort(cboCOMPort.SelectedItem.ToString(),
                                115200, Parity.None, 8, StopBits.One);
        _serialport.Handshake = Handshake.None;

        // Set the read/write timeouts
        _serialport.ReadTimeout = READ_TIMEOUT;
        _serialport.WriteTimeout = WRITE_TIMEOUT;
        _serialport.ReadBufferSize = REC_BUFFER_SIZE;
        _serialport.Open();

        _serialport.DataReceived +=
                    new SerialDataReceivedEventHandler(sp_DataReceived);

    }// end cboCOMPort_SelectedIndexChanged

}// end class

}// end namespace
```

Reference: *The handling of the USB port is based on an article by Ryan Alford (with added content by Arjun Walmiki, Gregory Krzywoszyja and Mahesh Chand) at: http://www.c-sharpcorner.com/uploadfile/eclipsed4utoo/communicating-with-serial-port-in-C-Sharp/*

At program start, the user first needs to select the applicable USB COM port, which initializes the port (*SelectedIndexChanged* event).

Beyond that, the program functions as a simple USB terminal: Messages are typed in the top text box and sent by clicking on the *Send* command button. The larger text box displays the received data.

The following shows a screen shot taken through a session of sending J1939 data:

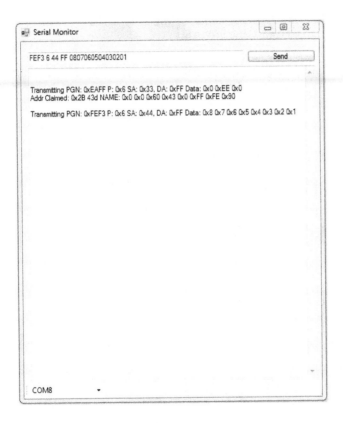

Note: *As usual, this software project is available through the download page at http://ard1939.com.*

3.2 ARD1939 - J1939 Protocol Stack for Arduino

The most interesting project in this book is, of course, the fully functional SAE J1939 protocol stack for the Arduino, the ARD1939. ARD1939 supports the full protocol, SAE J1939/21 (Network Management, Address Claim Process) and SAE J1939/81 (Transport Protocol, the transport of messages of up to 1785 bytes, including BAM and RTS/CTS sessions).

Please refer to chapter *The Two Elements of the SAE J1939 Protocol Stack* for a technical description of SAE J1939/21 and SAE J1939/81.

Note: *While a complete documentation of the SAE J1939 Standard is out of the scope of this book, I have addressed the mandatory basics in a previous chapter. For more detailed information, please refer to "A Comprehensible Guide to J1939" as listed in the appendix for recommended literature.*

3.2.1 ARD1939 - Function Overview

The programming of every serial communication (such as RS232, CAN, Ethernet, just to name a few) should follow a very simple sequence:

1. Initialization
2. Read data
3. Write data
4. Check status

These four function calls should be everything a programmer needs to access the serial protocol, and there is no reason why an SAE J1939 protocol stack implementation should not follow the same scheme. Digging inside complex code in order to understand how to access protocol functions is simply a waste of time.

Unfortunately, most commercially available protocol stacks shine through complexity and/or lack of documentation. To mention it upfront, the ARD1939 protocol stack requires only very few function calls and thus enables the programmer to put the protocol into use in a very short time.

The functions available to the SAE J1939 application layer (i.e. your program) are:

Initialization

- *j1939.Init* – Initializes the protocol stack settings
- *j1939.SetPreferredAddress* – Sets the preferred node (source) address
- *j1939.SetNAME* – Sets the ECU's NAME using the individual parameters
- *j1939.SetMessageFilter* – Sets the PGNs to be processed in your application

Read/Write – Check Status

- *j1939.Operate* – Handles the address claim process, reads PGNs from the vehicle network, and delivers the current protocol status (Address Claim in progress, Address Claim successful, Address Claim failed)
- *j1939.Transmit* – Transmits data to the vehicle network and handles the Transport Protocol (TP)

Other Application Functions

- *j1939.Terminate* – Resets the protocol stack settings
- *j1939.GetSourceAddress* – Delivers the negotiated node address
- *j1939.DeleteMessageFilter* – Deletes a message filter

For a detailed description of these function calls, please refer to the appendix "ARD1939 Protocol Stack Reference."

3.2.2 ARD1939 - Implementation

I provide the ARD1939 (SAE J1939 Protocol Stack for Arduino) "as is," i.e. I provide the entire project (Arduino Sketch). In order to copy and customize the project, use the "Save As" function in the Arduino's IDE.

I had contemplated creating an Arduino library, but I deem the assembly process (connection your project to the library, integrating the MCP2515 functionality and CAN interface functions, etc.) as too cumbersome. The complete project I provide will not only save you time, but it is also a proven solution and will thus not result in endless error corrections.

Note: *As a reminder, the Uno's restricted memory resources allow only a maximum J1939 data length of 256, while the Mega version supports the full 1785 bytes per message frame. Otherwise, both versions are identical in their functionality.*

3.2.2.1 Functionality Settings

Overall, the sketch supports three versions of the ARD1939 protocol stack, two for the Arduino Uno (ARD1939-Uno and ARD1939-Uno/TP) and one for the Arduino Mega 2560 (ARD1939-Mega).

1. **ARD1939-Uno** – For the sake of saving memory resources, this version does not support the Transport Protocol (TP) according to SAE J1939/21.
   Program Storage Space: ~8,200 bytes (~25%)
   Dynamic Memory: ~510 bytes (~25%)
   Message Filters: 10

2. **ARD1939-Uno/TP** – This version supports the Transport Protocol (TP) but only for message lengths up to 256 bytes.
   Program Storage Space: ~12,000 bytes (~37%)
   Dynamic Memory: ~1,100 bytes (~55%)
   Message Filters: 10

3. **ARD1939-Mega** – This version supports the full SAE J1939 protocol stack including Transport Protocol (TP) for message lengths up to 1785 bytes according to the J1939 standard.
   Program Storage Space: ~12,300 bytes (~4%)
   Dynamic Memory: ~4,600 bytes (~56%)

Note: *The exact program storage space and dynamic memory requirements also depend on the extend of the actual J1939 application in the loop () function.*

The functionality settings can be found inside the *ARD1939.h* file, which is part of the project.

```
// Program Version
// -------------------------------------------
// 0 - ARD1939-Uno
// 1 - ARD1939-Uno/TP
// 2 - ARD1939-Mega
#define ARD1939VERSION                        2

// J1939 Settings
#if ARD1939VERSION == 0
  #define TRANSPORT_PROTOCOL                  0
  #define J1939_MSGLEN                        8
  #define MSGFILTERS                          10
#endif

#if ARD1939VERSION == 1
  #define TRANSPORT_PROTOCOL                  1
  #define J1939_MSGLEN                        256
  #define MSGFILTERS                          10
#endif

#if ARD1939VERSION == 2
  #define TRANSPORT_PROTOCOL                  1
  #define J1939_MSGLEN                        1785
  #define MSGFILTERS                          100
#endif
```

The program version (ARD1939VERSION) allows the activation of the Transport Protocol (TP) where 1 = active and 0 = inactive. You can also experiment with the maximum message length (J1939_MSGLEN) and the number of supported message filters (MSGFILTERS), but please pay attention to the Arduino compiler message. It may warn you of memory size problems.

Also, feel free to browse through the file as it allows you a number of further customized settings such as:

- Preferred Source Address
- Address Range
- NAME settings

```
#define SA_PREFERRED                      128
#define ADDRESSRANGEBOTTOM                129
#define ADDRESSRANGETOP                   247

// NAME Fields Default
#define NAME_IDENTITY_NUMBER              0xFFFFFF
#define NAME_MANUFACTURER_CODE            0xFFF
#define NAME_FUNCTION_INSTANCE            0
#define NAME_ECU_INSTANCE                 0x00
#define NAME_FUNCTION                     0xFF
#define NAME_RESERVED                     0
#define NAME_VEHICLE_SYSTEM               0x7F
#define NAME_VEHICLE_SYSTEM_INSTANCE      0
#define NAME_INDUSTRY_GROUP               0x00
#define NAME_ARBITRARY_ADDRESS_CAPABLE    0x01
```

I am repeating myself here (see chapter *SAE J1939/81 – Address Claim Procedure*), but I deem the reference necessary at this point:

Note: *The NAME fields have been assigned in a way that they will not interfere when used within an existing vehicle network. This has been done by setting the Identity Number and Manufacturer Code to the maximum value, which will result in a more passive role during the address claim process. An ECU with a NAME of higher value is more likely to lose the competition with another node using the same address.*

All settings as shown are used for demonstration purposes only. In all consequence, you must follow the SAE's recommendations. Also, you alone (and not the author or publisher) are responsible for the final implementation and the results thereof.

3.2.2.2 Program Structure

As with every Arduino application, the initialization of data and the J1939 protocol takes place during the *setup()* function, while the actual application resides in *loop()*.

setup() - The *j1939.Init* function is mandatory to operate ARD1939. You also need to set the preferred address and a NAME (for test purposes you can use the project default settings). Setting an address range is optional.

ARD1939 will not deliver any data (PGNs) unless you set the message filter. ARD1939 supports up to 10 (UNO) or 100 (Mega 2560) message filters.

loop() – This function must start with a *delay()* call, directly followed by *j1939.Operate()*. It is mandatory that the *delay* time and the time passed to *j1939.Operate* are identical, otherwise the protocol timing will be off, which, in turn, may cause error conditions.

82

The following code sample demonstrates a sample J1939 protocol stack application with the ARD1939-specific functions highlighted:

```
void setup()
{
    // Set the serial interface baud rate
    Serial.begin(115200);

    // Initialize the J1939 protocol including CAN settings
    j1939.Init(SYSTEM_TIME);

    // Set up the supported PGNs
    j1939.SetMessageFilter(PGN_ComponentID);

    // Set the preferred address and address range
    j1939.SetPreferredAddress(128);
    j1939.SetAddressRange(128, 247);

    // Set the NAME
    j1939.SetNAME(0x00, 0xFFF, 0x00, 0x00, 0xFF, 0x7F, 0x00, 0x01, 0x00);

}// end setup

void loop()
{
    // Establish the timer base in units of milliseconds
    delay(SYSTEM_TIME);

    // Call the J1939 protocol stack
    nJ1939Status = j1939.Operate(&nMsgId, &lPGN, &pMsg[0], &nMsgLen,
                                 &nDestAddr, &nSrcAddr, &nPriority);

    // Check for reception of PGNs for our ECU/CA
    if(nMsgId == J1939_MSG_APP)
    {
        // Check J1939 protocol status
        switch(nJ1939Status)
        {
            case ADDRESSCLAIM_INPROGRESS;
                break;

            case NORMALDATATRAFFIC:
                // Determine the negotiated source address
                nAppAddress = j1939.GetSourceAddress();

                // Respond corresponding to received PGN
                switch(lPGN)
                {
                    // Check your PGNs

                }// end switch(lPGN)
                break;

            case ADDRESSCLAIM_FAILED:
```

```
                break;

           }// end switch(nJ1939Status)

      }// end if

}// end loop
```

Each Arduino sample application, as introduced in the following chapters, will follow this same scheme.

Note: *The ARD1939 sketch I provide through the download page also follows this template. I have, however, added some sample code that I used for the proof of concept. These sample code sections are commented out in the main file, but you can, of course, reactivate the sections you like to use.*

3.2.3 Proof of Concept

Naturally, with the impression that the Arduino hardware is a far cry away from commercially available, industrial solutions, some doubters may challenge even the point that the Arduino software is up to the task of running a full SAE J1939 protocol stack. Therefore, I deem it necessary to accomplish and document a basic proof of concept. And even if you consider such a proof unnecessary, take a look at the test examples as they shows details of node communication that are of interest for each programmer. These tests explain the actual SAE J1939 protocol in more detail than any other textbook or manual.

Note: *While I have tested the ARD1939 protocol characteristics in great detail, I have not documented all test results in this book, because not all of these scenarios provided a sufficient educational value.*

It is also important to note that, while my test environment records timestamps, the time values as shown on the various screen shots reflect the messages at the time when they reached the actual network, i.e. they do not necessarily reflect the ECU's internal timing.

Ironically, the Arduino is the perfect hardware to accomplish these tests, because it allows the quick setup of simulated malfunctions. Consequently, I used two Arduino systems, one as the test object and another one to simulate protocol malfunctions.

Specifically, I used an Arduino Mega 2560 version (ARD1939-Mega) as the test object, while an Arduino Uno system acted as the network simulator. In order to verify the address claim process plus the TP (Transport Protocol) communication process and timing, I modified the Arduino Uno's code to produce communication and timing errors. The test node (i.e. the Mega 2560), in turn, must detect the errors produced by the testing node.

Both nodes will have an initial address of 0x80 (128). Both nodes also use the same NAME settings with the exception of the *ECU Instance*. The Arduino Uno has set the *ECU Instance* in its NAME to '0', while the Arduino Mega uses an instance of '1'. This assures that the Uno will always win the address claim process against the Mega.

The settings in the code are:

1. ARD1939.h

```
// NAME Fields Default
#define NAME_IDENTITY_NUMBER              0xFFFFFF
#define NAME_MANUFACTURER_CODE            0xFFF
#define NAME_FUNCTION_INSTANCE            0
#define NAME_ECU_INSTANCE                 0x00 or 0x01
#define NAME_FUNCTION                     0xFF
#define NAME_RESERVED                     0
#define NAME_VEHICLE_SYSTEM               0x7F
#define NAME_VEHICLE_SYSTEM_INSTANCE      0
#define NAME_INDUSTRY_GROUP               0x00
#define NAME_ARBITRARY_ADDRESS_CAPABLE    0x01
```

2. ARD1939 – *setup()*

```
// Set the preferred address and address range
j1939.SetPreferredAddress(SA_PREFERRED);
j1939.SetAddressRange(ADDRESSRANGEBOTTOM, ADDRESSRANGETOP);

// Set the NAME
j1939.SetNAME(NAME_IDENTITY_NUMBER,
              NAME_MANUFACTURER_CODE,
              NAME_FUNCTION_INSTANCE,
              NAME_ECU_INSTANCE,
              NAME_FUNCTION,
              NAME_VEHICLE_SYSTEM,
              NAME_VEHICLE_SYSTEM_INSTANCE,
              NAME_INDUSTRY_GROUP,
              NAME_ARBITRARY_ADDRESS_CAPABLE);
```

Note: *Per SAE J1939/81 each ECU (or better Control Application, since an ECU can accommodate several CAs) must maintain a unique NAME.*

For the ultimate proof I used screen shots made with the ADFweb CAN analyzer software to document the communication between the two nodes, including timing (i.e. measuring response times and timeouts as defined by the SAE J1939 Standards Collection).

The following information can also very well serve as a generic SAE J1939 compliance test. However, there is no claim that the test methods as described are 100% complete or serve every test scenario.

3.2.3.1 Address Claim Procedure (SAE J1939/81)

The tested address claim functionality includes:

- Claiming address with no contending node
- Claiming address with contending node

3.2.3.1.1 Claiming Address With No Contending Node

The first and by far simplest test is to check whether or not the node (Arduino Mega 2560) engages into the address negotiation process. This test is accomplished by a simple power cycle (reset) of the unit when no other competing node (i.e. a node competing for the same address) is connected.

NR.	TIME	ID (HEX)	DATA (HEX)
1	1:11:07 AM.333.9	18EEFF80	FF FF FF FF 01 FF FE 80

The screen shot shows the *Address Claimed* message with a CAN message ID of 0x18EEFF80.

- 0x18 – Message priority = 6
- 0xEE – MSB of the *Address Claimed* PGN (60928 = 0xEE00)
- 0xFF – LSB of the *Address Claimed* PGN (destination address), indicating a broadcast message (using the global address of 255 as the destination address)
- 0x80 – Claimed address (source address)

In contrast let's have a look at the testing node (Arduino Uno) as it starts up under the same conditions.

NR	TIME	ID (HEX)	DATA (HEX)
1	1:21:04 AM.205.8	18EEFF80	FF FF FF FF 00 FF FE 80

The only difference here is the *ECU Instance* of 0x00 compared to the ox01 of the Arduino Mega node (5th byte from left in the DATA section).

3.2.3.1.2 Claiming Address With Contending Node

Now let's have the two nodes compete for their addresses, since they both use the same preferred source address. For that purpose, it is important to consider which node starts up first, and the test results will demonstrate the difference.

Test #1: Mega 2560 starts up first, then Uno

NR	TIME	ID (HEX)	DATA (HEX)
1	1:27:35 AM.123.7	18EEFF80	FF FF FF FF 01 FF FE 80
2	1:27:40 AM.445.7	18EEFF80	FF FF FF FF 00 FF FE 80
3	1:27:40 AM.446.8	18EEFF81	FF FF FF FF 01 FF FE 80

- **Line 1**: The Arduino Mega 2560 starts up and claims an address of 0x80 (128).
- **Line 2**: The Arduino Uno starts up and claims the same address.
- **Line 3**: The Arduino Mega 2560 had compared both NAMEs and found that the competing node (Uno) has a lower NAME (due to the ECU Instance). Consequently, the Mega 2560 claims the next available address of 0x81 (129).

Note: *Have a look at the ECU Instance (5th data byte) to identify the two nodes and their responses.*

Test #2: Uno starts up first, then Mega 2560

NR.	TIME	ID (HEX)	DATA (HEX)
1	1:37:00 AM.876.7	18EEFF80	FF FF FF FF 00 FF FE 80
2	1:37:07 AM.515.1	18EEFF80	FF FF FF FF 01 FF FE 80
3	1:37:07 AM.516.4	18EEFF80	FF FF FF FF 00 FF FE 80
4	1:37:07 AM.517.9	18EEFF81	FF FF FF FF 01 FF FE 80

This screen shot demonstrates clearly that the start-up sequence will result in a different communication process between the two nodes.

- **Line 1**: The Arduino Uno starts up and claims an address of 0x80 (128).

- **Line 2**: The Arduino Mega 2560 starts up and claims the same address.

- **Line 3**: The Arduino Uno verified both NAMEs and found that its NAME was lower. Consequently, it sends its *Address Claimed* message again.

- **Line 4**: The Arduino Mega 2560 had compared both NAMEs and found that the competing node (Uno) has a lower NAME (due to the ECU Instance). Consequently, the Mega 2560 claims the next available address of 0x81 (129).

The next test becomes even more interesting. In this case, I defined the Arduino Mega 2560 node as "not-arbitrary-address-capable," meaning the node will have only one negotiable address.

The code modifications are:

1. ARD1939.h

```
// NAME Fields Default
:
#define NAME_ARBITRARY_ADDRESS_CAPABLE          0x00
```

2. ARD1939 – *setup()*

```
// Set the preferred address and address range
j1939.SetPreferredAddress(SA_PREFERRED);
// j1939.SetAddressRange(ADDRESSRANGEBOTTOM, ADDRESSRANGETOP);
```

```
// Set the NAME
j1939.SetNAME(NAME_IDENTITY_NUMBER,
              NAME_MANUFACTURER_CODE,
              NAME_FUNCTION_INSTANCE,
              NAME_ECU_INSTANCE,
              NAME_FUNCTION,
              NAME_VEHICLE_SYSTEM,
              NAME_VEHICLE_SYSTEM_INSTANCE,
              NAME_INDUSTRY_GROUP,
              NAME_ARBITRARY_ADDRESS_CAPABLE);
```

Test #3: Mega 2560 starts up first, then Uno

NR	TIME	ID (HEX)	DATA (HEX)
1	1:58:11 AM.193.4	18EEFF80	FF FF FF FF 01 FF FE 00
2	1:58:16 AM.038.5	18EEFF80	FF FF FF FF 00 FF FE 80
3	1:58:16 AM.040.2	18EEFF80	FF FF FF FF 01 FF FE 00
4	1:58:16 AM.042.3	18EEFF81	FF FF FF FF 00 FF FE 80

The result is somewhat surprising (the Mega wins the address claim), but in the end it makes total sense.

- **Line 1**: The Arduino Mega 2560 starts up and claims an address of 0x80 (128).

- **Line 2**: The Arduino Uno starts up and claims the same address.

- **Line 3**: The Arduino Mega 2560 verified both NAMEs and found that its NAME was lower. Consequently, it sends its *Address Claimed* message again.

- **Line 4**: The Arduino Uno had compared both NAMEs and found that the competing node (Uno) has a lower NAME (not due to the ECU Instance, but the *Arbitrary Address Capable* flag). Consequently, the Uno claims the next available address of 0x81 (129).

Note: *The NAME, as it appears in the "Address Claimed" message, is LSB first and MSB last.*

Test #4: Uno starts up first, then Mega 2560

NR	TIME	ID (HEX)	DATA (HEX)
1	2:05:33 AM.304.7	18EEFF80	FF FF FF FF 00 FF FE 80
2	2:05:36 AM.691.5	18EEFF80	FF FF FF FF 01 FF FE 00
3	2:05:36 AM.693.0	18EEFF81	FF FF FF FF 00 FF FE 80

The result is obvious, and I will refrain here from explaining all details.

Note: *This last test scenario reveals one of the many ingenious details of the SAE J1939 protocol. Initially, one might assume that an ECU with a single source address has some disadvantages over those with multiple available addresses. However, the position of the "Arbitrary Address Capable" flag ensures that nodes "less capable" than others can still connect to the network.*

In a next step, we will create scenarios where the node is unable to claim a source address. For this purpose, we will set the Arbitrary Address Capable flag in both nodes to '0' and use the same single address of 0x80 (128).

Test #5: Mega 2560 starts up first, then Uno

NR	TIME	ID (HEX)	DATA (HEX)
1	12:11:23 AM.901.2	18EEFF80	FF FF FF FF 01 FF FE 00
2	12:11:23 AM.902.4	18EEFF80	FF FF FF FF 00 FF FE 00
3	12:11:23 AM.904.0	18EEFFFE	FF FF FF FF 01 FF FE 00

This is the first scenario where one of the two nodes, namely the Arduino Mega 2560, in the network is not able of claiming a source address.

- **Line 1**: The Arduino Mega 2560 starts up and claims a source address of 0x80 (128).

- **Line 2**: The Arduino Uno starts up and claims the same address.

- **Line 3**: The Arduino Mega 2560 compares both NAMEs and determines that the contending node has a higher priority (lower NAME). Since the node does not have any further addresses available, it sends out a *Cannot Claim Address* message.

Note: *The "Cannot Claim Address" message is identical to the "Address Claimed" message, but it reports the NULL address (254) as the source address.*

Test #6: Uno starts up first, then Mega 2560

NR	TIME	ID (HEX)	DATA (HEX)
1	1:00:09 AM.358.2	18EEFF80	FF FF FF FF 00 FF FE 00
2	1:00:10 AM.889.8	18EEFF80	FF FF FF FF 01 FF FE 00
3	1:00:10 AM.890.9	18EEFF80	FF FF FF FF 00 FF FE 00
4	1:00:10 AM.892.6	18EEFFFE	FF FF FF FF 01 FF FE 00

Here again (and as expected), the Arduino Uno wins the address claim process.

- **Line 1**: The Arduino Uno starts up and claims a source address of 0x80 (128).
- **Line 2**: The Arduino Mega 2560 starts up and claims the same address.
- **Line 3**: The Arduino has compared both NAMEs and found that its NAME has a higher priority (lower NAME). In consequence, it sends its *Address Claimed* message again.
- **Line 4**: The Arduino Mega 2560 compares both NAMEs and determines that the contending node has a higher priority (lower NAME). Since the node does not have any further addresses available, it sends out a *Cannot Claim Address* message.

Last, but not least, here comes the most interesting test, because it does involve some serious network simulation. The question is, does our J1939 ECU manage its available address range correctly?

For this purpose, we add code to the Arduino Uno's application program that will enable us to reject a defined set of addresses claimed by the Arduino Mega 2560.

The test conditions are:

- Both ECUs have a preferred source address of 0x80 (128).
- Both ECUs are *Arbitrary Address Capable*.
- Both EUCs support a negotiable address range between 129 and 135.
- The Arduino Uno has an ECU instance of 0x00.
- The Arduino Mega 2560 has an ECU instance of 0x01.

In both sketches, the Uno and the Mega 256, we change/add the following code in the *setup()* function (see highlighted sections):

```
// Set the preferred address and address range
j1939.SetPreferredAddress(SA_PREFERRED);
j1939.SetAddressRange(129, 135);
```

The addition to the Arduino Uno's *loop()* function is as shown here:

```
:
:
 byte msgFakeNAME[] = {0, 0, 0, 0, 0, 0, 0, 0};

// Establish the timer base in units of milliseconds
 delay(SYSTEM_TIME);

 // Call the J1939 protocol stack
 nJ1939Status = j1939.Operate(&nMsgId, &lPGN, &pMsg[0], &nMsgLen, &nDestAddr,
                              &nSrcAddr, &nPriority);

 // Block certain claimed addresses
 if(nMsgId == J1939_MSG_PROTOCOL)
 {
   if(lPGN == 0x00EE00)
   {
     if(nSrcAddr >= 129 && nSrcAddr <= 135)
       j1939.Transmit(6, 0x00EE00, nSrcAddr, 255, msgFakeNAME, 8);

   }// end if

 }// end if

 // Check for reception of PGNs for our ECU/CA
 if(nMsgId == J1939_MSG_APP)
 :
 :
```

First, we are using a "fake" NAME to assure that the Uno will win the process claim process no matter what. A NAME filled will all-zeros will do exactly that. After all, this is a simulation program.

Behind the *j1939.Operate(...)* function we are able to determine whether the received PGN is an application message (nMsgId = J1939_MSG_APP) or a network management message (nMsgId = J1939_MSG_PROTOCOL).

If the PGN is a network management message, we check whether or not it is the Address Claimed message. Further we check for the address range claimed by the other node.

Then follows the "nasty" line where we deny the Arduino Mega 2560 any address it attempts to claim by calling the *j1939.Transmit(...)* function. We transmit an *Address Claimed* message and use the same source address as claimed by the Mega 2560. Since the Uno's NAME is unbeatable, the Mega 2560 will try claiming more addresses until it runs out of resources.

The following screen shot shows exactly what I described:

NR	TIME	ID (HEX)	DATA (HEX)
1	1:54:55 AM.940.1	18EEFF80	FF FF FF FF 01 FF FE 80
2	1:54:55 AM.941.8	18EEFF80	FF FF FF FF 00 FF FE 80
3	1:54:55 AM.944.0	18EEFF81	FF FF FF FF 01 FF FE 80
4	1:54:55 AM.945.5	18EEFF81	00 00 00 00 00 00 00 00
5	1:54:55 AM.946.8	18EEFF82	FF FF FF FF 01 FF FE 80
6	1:54:55 AM.948.2	18EEFF82	00 00 00 00 00 00 00 00
7	1:54:55 AM.949.6	18EEFF83	FF FF FF FF 01 FF FE 80
8	1:54:55 AM.950.9	18EEFF83	00 00 00 00 00 00 00 00
9	1:54:55 AM.952.3	18EEFF84	FF FF FF FF 01 FF FE 80
10	1:54:55 AM.953.6	18EEFF84	00 00 00 00 00 00 00 00
11	1:54:55 AM.955.1	18EEFF85	FF FF FF FF 01 FF FE 80
12	1:54:55 AM.956.2	18EEFF85	00 00 00 00 00 00 00 00
13	1:54:55 AM.957.9	18EEFF86	FF FF FF FF 01 FF FE 80
14	1:54:55 AM.960.0	18EEFF86	00 00 00 00 00 00 00 00
15	1:54:55 AM.961.7	18EEFF87	FF FF FF FF 01 FF FE 80
16	1:54:55 AM.963.8	18EEFF87	00 00 00 00 00 00 00 00
17	1:54:55 AM.965.6	18EEFFFE	FF FF FF FF 01 FF FE 80

First the Arduino Mega 2560 attempts to claim its preferred address 0x80 (128), and, after being denied, attempts to claim 129 (0x81) through 135 (0x87).

Finally (line 17), it runs out of addresses and sends a *Cannot Claim Address* message.

As the final test, we will throw the Mega 2560 a bone and allow address 135 (0x87). That means, in the Uno's *loop()* function we change one parameter, i.e. we set the last address to 134 instead of 135.

```
// Block certain claimed addresses
if(nMsgId == J1939_MSG_PROTOCOL)
{
  if(lPGN == 0x00EE00)
  {
    if(nSrcAddr >= 129 && nSrcAddr <= 134)
      j1939.Transmit(6, 0x00EE00, nSrcAddr, 255, msgFakeNAME, 8);

  }// end if

}// end if
```

NR	TIME	ID (HEX)	DATA (HEX)
1	2:22:41 AM.080.2	18EEFF80	FF FF FF FF 01 FF FE 80
2	2:22:41 AM.081.8	18EEFF80	FF FF FF FF 00 FF FE 80
3	2:22:41 AM.083.0	18EEFF81	FF FF FF FF 01 FF FE 80
4	2:22:41 AM.084.5	18EEFF81	00 00 00 00 00 00 00 00
5	2:22:41 AM.085.7	18EEFF82	FF FF FF FF 01 FF FE 80
6	2:22:41 AM.087.2	18EEFF82	00 00 00 00 00 00 00 00
7	2:22:41 AM.088.5	18EEFF83	FF FF FF FF 01 FF FE 80
8	2:22:41 AM.089.8	18EEFF83	00 00 00 00 00 00 00 00
9	2:22:41 AM.091.3	18EEFF84	FF FF FF FF 01 FF FE 80
10	2:22:41 AM.092.5	18EEFF84	00 00 00 00 00 00 00 00
11	2:22:41 AM.094.1	18EEFF85	FF FF FF FF 01 FF FE 80
12	2:22:41 AM.095.2	18EEFF85	00 00 00 00 00 00 00 00
13	2:22:41 AM.096.8	18EEFF86	FF FF FF FF 01 FF FE 80
14	2:22:41 AM.098.9	18EEFF86	00 00 00 00 00 00 00 00
15	2:22:41 AM.100.7	18EEFF87	FF FF FF FF 01 FF FE 80

As the screen shot proves, the Mega 2560 goes through all available addresses until it gets to 135 (0x87). It is finally able to claim an address (line 15).

3.2.3.2 Sending and Receiving Messages

Proving the capability of the most basic task of sending and receiving regular 8-byte (CAN) messages must be seen as an unnecessary test, since that had already taken place in previous chapters.

Things change, however, with the transmission and reception of J1939 data frames with more than eight bytes.

94

Both sessions, BAM for messages broadcasting and RTS/CTS for peer-to-peer communication, are based on the exchange of multiple data packages and they require exact timing controls. In other words, both protocol functions come with a number of timeouts that need to be observed.

Note: *The SAE J1939/21 Standard requires that an ECU supports one BAM and one RTS/CTS session simultaneously, and the ARD1939 protocol stack fulfills that requirement.*

3.2.3.3 Transport Protocol – BAM Session (SAE J1939/21)

Test items include:

- Basic function

- Timing control

The initial test of the BAM session requires only one ECU, in this case I am using the full protocol running on the Arduino Mega 2560 (The Arduino Uno is still connected to the network, thus the Mega 2560 will claim a source address of 0x81 = 129).

The code we added to the Mega 2560's *loop()* function is highlighted:

```
      :
      :
// Variables for proof of concept tests
byte msgLong[] = {0x31, 0x32, 0x33, 0x34, 0x35, 0x36, 0x37, 0x38, 0x39, 0x40,
                  0x41, 0x42, 0x43, 0x44, 0x45};

// Establish the timer base in units of milliseconds
delay(SYSTEM_TIME);

// Call the J1939 protocol stack
nJ1939Status = j1939.Operate(&nMsgId, &lPGN, &pMsg[0], &nMsgLen, &nDestAddr,
                             &nSrcAddr, &nPriority);

// Send out a periodic message with a length of more than 8 bytes
// BAM Session
if(nJ1939Status == NORMALDATATRAFFIC)
{
  nCounter++;

  if(nCounter == (int)(5000/SYSTEM_TIME))
  {
    nSrcAddr = j1939.GetSourceAddress();
    j1939.Transmit(6, 59999, nSrcAddr, 255, msgLong, 15);
    nCounter = 0;

  }// end if

}// end if
```

First, we use a global variable nCounter, which is set to its default value of zero.

Inside the *loop()* function we create a "long" message with 15 data bytes. Then we check the address claim status, and we continue when "normal data traffic" was reached, meaning the address claim process was successful.

Every five seconds (randomly chosen) we determine the acquired source address, and send a PGN (59999, also randomly chosen) using the long message. We also use a destination address of 255, the global address, which should trigger a BAM session.

Let's have a look at the screen shot made from the ADFweb CAN analyzer software tool:

NR	TIME	ID (HEX)	DATA (HEX)
1	12:13:56 AM.942.4	1CECFF81	20 0F 00 03 FF 5F EA 00
2	12:13:56 AM.996.9	1CEBFF81	01 31 32 33 34 35 36 37
3	12:13:57 AM.051.5	1CEBFF81	02 38 39 40 41 42 43 44
4	12:13:57 AM.106.1	1CEBFF81	03 45 FF FF FF FF FF FF

Line 1:
- The ECU sends the BAM (Broadcast Announce Message).

- **ID (HEX):**
- 0x1C indicates a priority of 7.
- 0xECFF initiates the TP (Transport Protocol), meaning this is the TP.CM (Connection Management) PGN of 0xEC00 (60416) plus the global address of 0xFF (255).
- 0x81 (129) is the ECU's source address

- **DATA (HEX):**
- 0x20 = 32 = Control byte indicating a BAM
- 0x000F indicates a total length of 15 data bytes (LSB first, MSB last)
- 0x03 indicates the number of packets to be transmitted
- 0xFF (Reserved, according to SAE J1939/21)
- 0x00EA5F indicates the PGN (59999) of the multi-packet message (LSB first, MSB last)

Line 2 through Line 4:

- Data sent by the ECU at address 0x81 (129)

- The first byte indicates the sequence number, in this case going from 1 to 3

- The following 7 bytes are raw data.

- All unused data in the last package is being set to 0xFF

3.2.3.3.1 Message Timing

NR	TIME	ID (HEX)	DATA (HEX)
1	12:13:56 AM.942.4	1CECFF81	20 0F 00 03 FF 5F EA 00
2	12:13:56 AM.996.9	1CEBFF81	01 31 32 33 34 35 36 37
3	12:13:57 AM.051.5	1CEBFF81	02 38 39 40 41 42 43 44
4	12:13:57 AM.106.1	1CEBFF81	03 45 FF FF FF FF FF FF

Note the time stamps from line 1 through line 4 of the screen shot (look at the last four numbers indicating the time in tenth of milliseconds; for instance 942.4 milliseconds in line 1). SAE J1939/21 requires a packet frequency between 50 to 200 milliseconds, which is being met by the Mega 2560 ECU.

A timeout will occur when a time of greater than 750 milliseconds elapsed between two message packages when more packets were expected. A timeout will close the connection, but there will be no message exchange indicating a detected timeout.

Note: *In this previous example, the Arduino Uno played only a minor role in the J1939 network. In all consequence, with the ARD1939-Uno/TP protocol stack installed, it will receive the BAM message. However, in order for the message to reach the actual application, you need to set the filter to allow the PGN being passed to the application.*

3.2.3.4 Transport Protocol – RTS/CTS Session (SAE J1939/21)

Test items include:

- Basic function

- Timing control

Naturally, for the RTS/CTS session we will need two SAE J1939 nodes. In my setup that means communication between the Arduino Uno and the Mega 2560 ECUs.

The test conditions are similar to the previous examples:

- The SAE J1939 protocol stack is installed on both systems.

- The Arduino Uno has a preferred address of 0x80 (128) and an ECU instance of '0'.

- The Arduino Mega has a preferred address of 0x80 (128) and an ECU instance of '1', which means, after the address claim process, the Mega 2560 will get 0x81 (129) as its source address.

- The Arduino Mega will send the 15-byte message (as copied from the previous example) to the Arduino Uno.

The code changes in the Mega 2560's *loop()* function are minimal:

```
    :
    :
// Variables for proof of concept tests
byte msgLong[] = {0x31, 0x32, 0x33, 0x34, 0x35, 0x36, 0x37, 0x38, 0x39, 0x40,
                  0x41, 0x42, 0x43, 0x44, 0x45};

// Establish the timer base in units of milliseconds
delay(SYSTEM_TIME);

// Call the J1939 protocol stack
nJ1939Status = j1939.Operate(&nMsgId, &lPGN, &pMsg[0], &nMsgLen, &nDestAddr,
                             &nSrcAddr, &nPriority);

// Send out a periodic message with a length of more than 8 bytes
// RTS/CTS Session
if(nJ1939Status == NORMALDATATRAFFIC)
{
  nCounter++;

  if(nCounter == (int)(5000/SYSTEM_TIME))
  {
    nSrcAddr = j1939.GetSourceAddress();
    j1939.Transmit(6, 59999, nSrcAddr, 0x80, msgLong, 15);
    nCounter = 0;

  }// end if

}// end if
    :
    :
```

Instead of sending the message to the global address (255), we will send it to the Arduino Uno's address of 0x80.

We also need to change some code in the <u>Arduino Uno's</u> *setup()* function, i.e. we need to allow the transmitted PGN by setting the message filter. The (highlighted) code we insert looks as follows:

```
// Set the preferred address and address range
j1939.SetPreferredAddress(SA_PREFERRED);
j1939.SetAddressRange(ADDRESSRANGEBOTTOM, ADDRESSRANGETOP);

// Set the message filter
j1939.SetMessageFilter(59999);

// Set the NAME
j1939.SetNAME(NAME_IDENTITY_NUMBER,
              NAME_MANUFACTURER_CODE,
              NAME_FUNCTION_INSTANCE,
              NAME_ECU_INSTANCE,
              NAME_FUNCTION,
              NAME_VEHICLE_SYSTEM,
              NAME_VEHICLE_SYSTEM_INSTANCE,
              NAME_INDUSTRY_GROUP,
              NAME_ARBITRARY_ADDRESS_CAPABLE);
```

If we don't set the message filter, we will see the following communication on the bus:

NR	TIME	ID (HEX)	DATA (HEX)
1	2:18:19 AM.814.8	1CEC8081	10 0F 00 03 FF 5F EA 00
2	2:18:19 AM.816.8	1CEC8180	FF 02 FF FF FF 5F EA 00

- **Line 1**: The Mega 2560 transmits the Request to Send message (source address = 0x80, destination address 0x81)
- **Line 2**: The Arduino Uno responds with a Connection Abort message (note the first data byte = 255) and states that it's lacking the necessary resources (see second data byte = 0x02).

Note: *The SAE J1939/21 Standard does not address all possible scenarios that may lead to a "Connection Abort" message. I deemed it necessary to provide a feedback when an RTS/CTS session cannot proceed. In this case, the reason is that the node has no use for the message.*

Things get really interesting when the Arduino Uno has set the filter to allow the PGN:

NR	TIME	ID (HEX)	DATA (HEX)
1	2:43:00 AM.295.2	1CEC8081	10 0F 00 03 FF 5F EA 00
2	2:43:00 AM.296.3	1CEC8180	11 03 01 FF FF 5F EA 00
3	2:43:00 AM.352.4	1CEB8081	01 31 32 33 34 35 36 37
4	2:43:00 AM.407.1	1CEB8081	02 38 39 40 41 42 43 44
5	2:43:00 AM.461.8	1CEB8081	03 45 FF FF FF FF FF FF
6	2:43:00 AM.464.0	1CEC8180	13 0F 00 03 FF 5F EA 00

Line 1:
- The ECU at address 0x81 (129) sends a *Request to Send* message to the ECU at address 0x80 (128).

- **ID (HEX):**
- 0x1C indicates a priority of 7.

- 0xEC80 initiates the TP (Transport Protocol), meaning this is the TP.CM (Connection Management) PGN of 0xEC00 (60416) plus the destination address of 0x80 (128).

- 0x81 (129) is the ECU's source address

- **DATA (HEX):**
- 0x10 = 16 = Control byte indicating a *Request to Send* message

- 0x000F indicates a total length of 15 data bytes (LSB first, MSB last)

- 0x03 indicates the number of packets to be transmitted

- 0xFF (Reserved, according to SAE J1939/21)

- 0x00EA5F indicates the PGN (59999) of the multi-packet message (LSB first, MSB last)

Line 2:
- The ECU at address 0x80 (128) sends a *Clear to Send* message to the ECU at address 0x81 (129)

- **ID (HEX):**
- 0x1C indicates a priority of 7.
- 0xEC81 initiates the TP (Transport Protocol), meaning this is the TP.CM (Connection Management) PGN of 0xEC00 (60416) plus the destination address of 0x81 (129).
- 0x80 (128) is the ECU's source address
- **DATA (HEX):**
- 0x11 = 17 = Control byte indicating a *Clear to Send* message
- 0x03 indicates the number of packets to be transmitted
- 0x01 indicates the next packet number
- The next two bytes are reserved (0xFF)
- 0x00EA5F indicates the PGN (59999) of the multi-packet message (LSB first, MSB last)

Line 3 through Line 5:
- Data sent by the ECU at address 0x81 (129) to the ECU at address 0x80 (128)
- The first byte indicates the sequence number, in this case going from 1 to 3
- The following 7 bytes are raw data.
- All unused data in the last package is being set to 0xFF

Line 6:
- The ECU at address 0x80 (128) sends an *End of Message Acknowledgment* message to the ECU at address 0x81 (129)

- **ID (HEX):**
- 0x1C indicates a priority of 7.
- 0xEC81 initiates the TP (Transport Protocol), meaning this is the TP.CM (Connection Management) PGN of 0xEC00 (60416) plus the destination address of 0x81 (129).
- 0x80 (128) is the ECU's source address

- **DATA (HEX):**
- 0x13 = 19 = Control byte indicating a *End of Message Acknowledgment* message
- 0x000F indicates the message length (LSB first, MSB last)
- 0x03 indicates the total number of packages
- The next byte is reserved (0xFF)
- 0x00EA5F indicates the PGN (59999) of the multi-packet message (LSB first, MSB last)

3.2.3.4.1 Message Timing

NR	TIME	ID (HEX)	DATA (HEX)
1	2:43:00 AM.295.2	1CEC8081	10 0F 00 03 FF 5F EA 00
2	2:43:00 AM.296.3	1CEC8180	11 03 01 FF FF 5F EA 00
3	2:43:00 AM.352.4	1CEB8081	01 31 32 33 34 35 36 37
4	2:43:00 AM.407.1	1CEB8081	02 38 39 40 41 42 43 44
5	2:43:00 AM.461.8	1CEB8081	03 45 FF FF FF FF FF FF
6	2:43:00 AM.464.0	1CEC8180	13 0F 00 03 FF 5F EA 00

Note the time stamps from line 3 through line 5 of the screen shot (look at the last four numbers indicating the time in tenth of milliseconds; for instance 352.4 milliseconds in line 3). SAE J1939/21 requires a packet frequency between 50 to 200 milliseconds, which is being met by the Mega 2560 ECU.

3.2.3.4.2 Clear to Send Timeout

The simplest test for a *Clear to Send* timeout is accomplished by running the Mega 2560 ECU as the only node in the network.

To test this scenario, I run the Mega 2560 at an address of 0x80 (128) and attempt to initiate an RTS/CTS session with a node at address 0x33 (51).

NR	TIME	ID (HEX)	DATA (HEX)
1	1:56:31 AM.050.7	18EEFF80	FF FF FF FF 01 FF FE 80
2	1:56:36 AM.743.8	1CEC3380	10 0F 00 03 FF 5F EA 00
3	1:56:36 AM.961.6	1CEC3380	FF 03 FF FF FF 5F EA 00

- **Line 1**: The Arduino Mega 2560 claims the address 0x80.

- **Line 2**: It sends the *Request to Send* message to the node at address 0x33.

- **Line 3**: After a little over 200 milliseconds, it transmits the *Connection Abort* message, citing a timeout as the abort reason.

3.2.3.4.3 End of Message Acknowledgment Timeout

In order to test all timeouts during an RTS/CTS session, I used a dedicated Arduino sketch to simulate all the "nasty" features that would trigger a communication breakdown. This method was easier to implement rather than using a full SAE J1939 stack. Fiddling with the protocol stack code to simulate error conditions turned out to be time-consuming and it brought no improvement to the actual code (actually, quite the opposite, since it would blow up the code size).

As with all previous examples, this Arduino sketch (implemented on the Uno) is quite simple, and it leaves you all the possibilities of simulating a defective J1939 node.

First, we need to define the *Clear to Send* message in the global memory space:

- `byte msg_CTS[] = {0x11, 0x03, 0x01, 0xFF, 0xFF, 0x5F, 0xEA, 0x00};`

The data is based on the previously described RTS/CTS session (see the DATA section in line 2).

NR	TIME	ID (HEX)	DATA (HEX)
1	2:43:00 AM.295.2	1CEC8081	10 0F 00 03 FF 5F EA 00
2	2:43:00 AM.296.3	1CEC8180	11 03 01 FF FF 5F EA 00
3	2:43:00 AM.352.4	1CEB8081	01 31 32 33 34 35 36 37
4	2:43:00 AM.407.1	1CEB8081	02 38 39 40 41 42 43 44
5	2:43:00 AM.461.8	1CEB8081	03 45 FF FF FF FF FF FF
6	2:43:00 AM.464.0	1CEC8180	13 0F 00 03 FF 5F EA 00

The Arduino sketch, i.e. the *loop()* function was tailored for this specific RTS/CTS session:

```
void loop()
{
  // Declarations
  byte nPriority = 0;
  long lPGN = 0;
  byte nSrcAddr = 0;
  byte nDestAddr = 0;
  byte nData[8];
  int nDataLen;

  // Check for received J1939 messages
  if(j1939Receive(&lPGN, &nPriority, &nSrcAddr, &nDestAddr, nData, &nDataLen)
    == 0)
  {
    switch(lPGN)
    {
      case 0xEC00:  // RTS/CTS Session PGN

        // Check for CTS format
        switch(nData[0])
        {
          case 0x10:  // Request to Send

            // Send Clear to Send
            j1939Transmit(0xEC00, 7, 0x33, 0x80, &msg_CTS[0], 8);
            Serial.print("Request to Send.\n\r");
            break;

          case 0xFF:  // Connection Abort
            Serial.print("Connection Abort.\n\r");
            break;

        }// end switch

        break;

      default:
        break;

    }// end switch

  }// end if

}// end loop
```

As soon as the *j1939Receive(...)* function returns a valid data frame, we verify the PGN. If the PGN is the Transport Protocol (TP) PGN, we look at the first byte in the data array to determine if the message is a *Request to Send* or a *Connection Abort* message.

If it is a *Request to Send*, the Arduino transmits the *Clear to Send* data frame. However, the sketch will not send its *End of Message Acknowledgment* and that should trigger a *Connection Abort* message from the Mega 2560.

NR	TIME	ID (HEX)	DATA (HEX)
1	2:23:58 AM.081.7	1CEC3380	10 0F 00 03 FF 5F EA 00
2	2:23:58 AM.082.8	1CEC8033	11 03 01 FF FF 5F EA 00
3	2:23:58 AM.138.8	1CEB3380	01 31 32 33 34 35 36 37
4	2:23:58 AM.193.5	1CEB3380	02 38 39 40 41 42 43 44
5	2:23:58 AM.248.2	1CEB3380	03 45 FF FF FF FF FF FF
6	2:23:59 AM.608.5	1CEC3380	FF 03 FF FF FF 5F EA 00

- **Line 1**: The Mega 2560 sends the *Request to Send* message.

- **Line 2**: The Uno transmits the *Clear to Send* message.

- **Line 3 – 5**: The Mega transmits the data.

- **Line 6**: After a timeout without receiving the *End of Message Acknowledgment*, the Mega sends out the *Connection Abort* message citing a timeout.

3.2.3.4.4 RTS/CTS Session Test Program

Just for kicks, I extended the previously used Arduino sketch to simulate a fully functional RTS/CTS session (yet again, the code is highly tailored to the previous RTS/CTS session). Based on that sketch, I could simulate all error scenarios on the receiver's side.

NR	TIME	ID (HEX)	DATA (HEX)
1	3:13:52 AM.814.9	1CEC3380	10 0F 00 03 FF 5F EA 00
2	3:13:52 AM.815.9	1CEC8033	11 03 01 FF FF 5F EA 00
3	3:13:52 AM.872.0	1CEB3380	01 31 32 33 34 35 36 37
4	3:13:52 AM.926.7	1CEB3380	02 38 39 40 41 42 43 44
5	3:13:52 AM.981.4	1CEB3380	03 45 FF FF FF FF FF FF
6	3:13:52 AM.982.7	1CEC8033	13 0F 00 03 FF 5F EA 00

This screen shot shows the complete RTS/CTS session with the Arduino Uno sketch acting as the simulated communication partner.

Without explaining every single line (that has been done before), the complete RTS/CTS session took place according to SAE J1939/81.

The modified code is as follows (see highlighted sections):

Global memory space:

```
byte msg_CTS[] = {0x11, 0x03, 0x01, 0xFF, 0xFF, 0x5F, 0xEA, 0x00};
byte msg_ACK[] = {0x13, 0x0F, 0x00, 0x03, 0xFF, 0x5F, 0xEA, 0x00};
int nPackages = 0;
int nCounter = 0;
```

loop() function:

```
void loop()
{
  // Declarations
  byte nPriority = 0;
  long lPGN = 0;
  byte nSrcAddr = 0;
  byte nDestAddr = 0;
  byte nData[8];
  int nDataLen;

  // Check for received J1939 messages
  if(j1939Receive(&lPGN, &nPriority, &nSrcAddr, &nDestAddr, nData, &nDataLen)
     == 0)
  {
    switch(lPGN)
    {
      case 0xEC00:  // RTS/CTS Session PGN

        // Check for CTS format
        switch(nData[0])
        {
          case 0x10:  // Request to Send

            // Send Clear to Send
            j1939Transmit(0xEC00, 7, 0x33, 0x80, &msg_CTS[0], 8);
            Serial.print("Request to Send.\n\r");

            // Remember the number of packages
            nPackages = (int)nData[3];

            break;

          case 0xFF:  // Connection Abort
            Serial.print("Connection Abort.\n\r");
            break;

        }// end switch

        break;
```

106

```
     case 0xEB00: // RTS/CTS Data Packet

       nCounter++;
       Serial.print("Packet: ");
       Serial.print(nCounter);
       Serial.print(" of ");
       Serial.print(nPackages);
       Serial.print("\n\r");

       if(nCounter == nPackages)
       {
         // Last data packet arrived
         j1939Transmit(0xEC00, 7, 0x33, 0x80, &msg_ACK[0], 8);

         Serial.print("Sending ACK.\n\r");
         nCounter = 0;

       }// end if

     default:
       break;

   }// end switch

 }// end if

}// end loop
```

Note: *As always, this software project is available through the download page at http://ard1939.com.*

3.2.4 ARD1939 - Sample Application

The typical SAE J1939 ECU application not only involves the protocol stack but also a good amount of input and/or output processing. This could involve reading sensors and sending the result in form of a PGN or reading a PGN and setting an output (digital or analog).

However, since my focus is on the J1939 protocol stack and not on Arduino I/O programming, I will not go into the details of accessing the Arduino's hardware interfaces. There are a myriad of good books available that more than sufficiently cover this particular topic (see also my recommendation in the appendix for recommended literature).

Nevertheless, I have included a brief sample into the ARD1939 project to process queries for:

- PGN 65242 – Software Identification
- PGN 65259 – Component Identification
- PGN 65267 – Vehicle Position

First, let's have a look at these PGNs:

- **PGN 65242** – Software Identification
    - o Transmission Rate: On Request
    - o Data Length: Variable
    - o Default Priority: 6
    - o PG Number: 65242 (FEDAhex)
    - o **Data:**
        - Byte 1: Number of Software ID fields (user-specific)
        - Bytes 2…n: Software Identification

- **PGN 65259** – Component Identification
    - o Transmission Rate: On Request
    - o Data Length: Variable
    - o Default Priority: 6
    - o PG Number: 65259 (FEEBhex)
    - o **Data:**
        - Byte 1 – 5: Make
        - Variable, up to 200 characters: Model
        - Variable, up to 200 characters: Serial Number
        - Variable, up to 200 characters: Unit Number

Both messages, software and component identification can contain several (optional) fields as described, which must be separated by an "*" (delimiter). It is not necessary to include all fields, however, the "*" delimiter is always required.

- **PGN 65267** – Vehicle Position
  - Transmission Rate: 5 sec
  - Data Length: 8 bytes
  - Default Priority: 6
  - PG Number: 65267 (FEF3hex)
  - **Data:**
    - Byte 1 – 4: Latitude
    - Byte 5 – 8: Longitude

In terms of programming these PGNs into our application, we first need to make up the message's content (in this case, for demo purposes only) and define the PGNs.

In the global memory space (above *setup()*) we add the following:

```
// PGNs and messages
#define PGN_RequestMessage      0xEA00

#define PGN_SoftwareID          65242
#define PGN_ComponentID         65259
#define PGN_VehiclePosition     65267

#define MSGLEN_SoftwareID       12
#define MSGLEN_ComponentID      33
#define MSGLEN_VehiclePosition  8

byte msgSoftwareID[] = {" V*1*00*01**"};
byte msgComponentID[] = {"MAKEX*MODEL_123*01234567890-007**"};
byte msgVehiclePosition[] = {0x08, 0x07, 0x06, 0x05, 0x04, 0x03, 0x02, 0x01};
```

Note: *The content of all messages was randomly chosen.*

In the *setup()* routine, we need to define the various message filter, in this case only for the request message (Software ID, Component ID, and vehicle position are not received; they are responses from our J1939 application).

```
// Set the message filter
j1939.SetMessageFilter(PGN_RequestMessage);
```

We leave all other settings such as preferred source address, address range, and NAME to their default values (see file ARD1939.h).

It is <u>important to know</u> that there is a specific ARD1939 feature that involves the handling of the *Request* message:

As defined in the SAE J1939 Standard, the *Request* message is 0xEA00, where the LSB is used as the destination address, i.e. the address of the ECU that is supposed to provide the requested information (e.g. 0xEA80, where 0x80 is the destination address).

The ARD1939 protocol stack's message filter, however, will allow <u>every</u> message in the 0xEAxx range to pass when you set any filter PGN in the 0xEAxx range.

This behavior is necessary to allow the global address (255), meaning there are scenarios where one ECU requests the information from all nodes in the network.

As a consequence, it is mandatory that the application, in addition to the *Request* message, also verifies the destination address with its own address.

In the *loop()* function, we add the code to check for the *Request* message and the requested PGN as highlighted in the following:

```
// Call the J1939 protocol stack
nJ1939Status = j1939.Operate(&nMsgId, &lPGN, &pMsg[0], &nMsgLen, &nDestAddr,
                             &nSrcAddr, &nPriority);

// Check for reception of PGNs for our ECU/CA
if(nMsgId == J1939_MSG_APP)
{
  // Check J1939 protocol status
  switch(nJ1939Status)
  {
    case ADDRESSCLAIM_INPROGRESS:

      break;

    case NORMALDATATRAFFIC:

      // Determine the negotiated source address
      byte nAppAddress;
      nAppAddress = j1939.GetSourceAddress();

      // Respond corresponding to received PGN
      switch(lPGN)
      {
        case PGN_RequestMessage:

          // Verify that we are the provider of the requested information
          if(nDestAddr == GLOBALADDRESS
          || nDestAddr == nAppAddress)
```

```
  {
    // Determine the requested PGN; note that LSB comes first
    long lRequestedPGN = ((long)pMsg[2] << 16)
                       + ((long)pMsg[1] << 8)
                       + (long)pMsg[0];

    // NOTE: In the following, nSrcAddr is the address of the
    //       requesting node.
    switch(lRequestedPGN)
    {
      case PGN_SoftwareID:
        Serial.print("Sending Software Identification.\n\r");

        // Add the number of fields to the SW ID message
        msgSoftwareID[0] = 4;

        j1939.Transmit(6, PGN_SoftwareID, nAppAddress, nSrcAddr,
                       &msgSoftwareID[0], MSGLEN_SoftwareID);
        break;

      case PGN_ComponentID:
        Serial.print("Sending Component Identification.\n\r");
        j1939.Transmit(6, PGN_ComponentID, nAppAddress, nSrcAddr,
                       &msgComponentID[0], MSGLEN_ComponentID);
        break;

      case PGN_VehiclePosition:
        Serial.print("Sending Vehicle Position.\n\r");
        j1939.Transmit(6, PGN_VehiclePosition, nAppAddress, nSrcAddr,
                       &msgVehiclePosition[0],
                       MSGLEN_VehiclePosition);
        break;

    }// end switch

    }// end if

    break;

    }// end switch(lPGN)

    break;

  case ADDRESSCLAIM_FAILED:

    break;

  }// end switch(nJ1939Status)

}// end if
```

Note: *The added code for this J1939 application will work on both Arduino hardware versions, the Uno and the Mega 2560.*

In order to test the application, I did not use another Arduino this time but entered the data for the *Request* message manually into the ADFweb software tool as shown here:

NR	ID(HEX)	DESCRIPTION	DATA (HEX)
1	18EA8033	Request for Software ID	DAFE00
2	18EA8033	Request for Component ID	EBFE00
3	18EA8033	Request for Vehicle Position	F3FE00

- **Line 1**: Node 0x33 requests the Software Identification from node 0x80.

- **Line 2**: Node 0x33 requests the Component Identification from node 0x80.

- **Line 3**: Node 0x33 requests the Vehicle Position from node 0x80.

Note: *Since the Arduino is the only J1939 node in the network, it will claim address 0x80 (128). In the ADFweb software tool, I used 0x33 as the requestor's address.*

The printout on the Arduino's serial monitor is as expected:

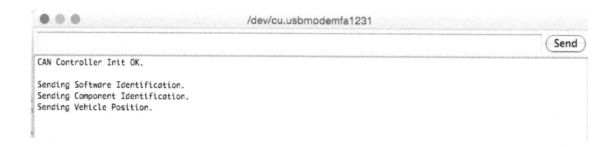

However, the J1939 data traffic as recorded on the bus came initially as a surprise, but, naturally, then it all made sense.

NR	TIME	ID (HEX)	DATA (HEX)
1	12:18:29 AM.851.0	18EEFF80	FF FF FF FF 01 FF FE 80
2	12:18:48 AM.757.0	18EA8033	DA FE 00
3	12:18:48 AM.759.7	1CEC3380	10 0C 00 02 FF DA FE 00
4	12:18:48 AM.976.0	1CEC3380	FF 03 FF FF FF DA FE 00
5	12:18:53 AM.274.0	18EA8033	EB FE 00
6	12:18:53 AM.276.9	1CEC3380	10 21 00 05 FF EB FE 00
7	12:18:53 AM.493.2	1CEC3380	FF 03 FF FF FF EB FE 00
8	12:18:59 AM.220.6	18EA8033	F3 FE 00
9	12:18:59 AM.222.8	18FEF380	08 07 06 05 04 03 02 01

- **Line 1**: The Arduino claims address 0x80.
- **Line 2**: *Request* message for software identification.
- **Line 3**: The Arduino sends a *Request to Send* message.
- **Line 4**: The Arduino sends a *Connection Abort* message due to timeout.
- **Line 5**: *Request* message for component identification.
- **Line 6**: The Arduino sends a *Request to Send* message.
- **Line 7**: The Arduino sends a *Connection Abort* message due to timeout.
- **Line 8**: *Request* message for Vehicle Position.
- **Line 9**: The Arduino sends the Vehicle Position data.

First of all, lines 8 and 9 bare no surprise. The PGN for Vehicle Position (65267 = 0xFEF3) does not fall into the address range for peer-to-peer communication, thus the message ID contains only the senders address (0x80), i.e. that of the Arduino.

However, in the case of message requests for Software and Component Identification, the ARD1939 reports a timeout through means of the *Connection Abort* message. Both messages, software and component identification, are longer than eight bytes and thus require the use of the Transportation Protocol (TP), in this case an RTS/CTS session, since this is a peer-to-peer communication. The timeout is caused, because there is no J1939 node with the address 0x33 in the network (and the ADFweb tool acts only as a passive network member).

In order to establish a communication without timeout errors, we need to connect another Arduino node with a source address of 0x33, and in the *setup()* routine we need to set filters for all three PGNs, software identification (65242), component identification (65259), and vehicle position (65267).

Finally, the result is as expected:

NR	TIME	ID (HEX)	DATA (HEX)
1	12:22:47 AM.491.9	18EEFF33	FF FF FF FF 00 FF FE 80
2	12:24:22 AM.154.6	18EEFF80	FF FF FF FF 01 FF FE 80
3	12:24:44 AM.399.3	18EA8033	DA FE 00
4	12:24:44 AM.402.4	1CEC3380	10 0C 00 02 FF DA FE 00
5	12:24:44 AM.404.5	1CEC8033	11 02 01 FF FF DA FE 00
6	12:24:44 AM.460.2	1CEB3380	01 04 56 2A 31 2A 30 30
7	12:24:44 AM.514.5	1CEB3380	02 2A 30 31 2A 2A FF FF
8	12:24:44 AM.516.0	1CEC8033	13 0C 00 02 FF DA FE 00
9	12:24:55 AM.763.2	18EA8033	EB FE 00
10	12:24:55 AM.766.3	1CEC3380	10 21 00 05 FF EB FE 00
11	12:24:55 AM.768.2	1CEC8033	11 05 01 FF FF EB FE 00
12	12:24:55 AM.824.1	1CEB3380	01 4D 41 4B 45 58 2A 4D
13	12:24:55 AM.878.4	1CEB3380	02 4F 44 45 4C 5F 31 32
14	12:24:55 AM.932.7	1CEB3380	03 33 2A 30 31 32 33 34
15	12:24:55 AM.987.0	1CEB3380	04 35 36 37 38 39 30 2D
16	12:24:56 AM.041.3	1CEB3380	05 30 30 37 2A 2A FF FF
17	12:24:56 AM.042.8	1CEC8033	13 21 00 05 FF EB FE 00
18	12:25:04 AM.589.0	18EA8033	F3 FE 00
19	12:25:04 AM.591.2	18FEF380	08 07 06 05 04 03 02 01

- **Line 1**: One Arduino claims address 0x33.
- **Line 2**: The other Arduino claims address 0x80.
- **Line 3**: *Request* message for software identification.
- **Line 4 – 8**: Software identification is transmitted and confirmed.
- **Line 9**: *Request* message for component identification.
- **Line 10 – 17**: Component identification is transmitted and confirmed.
- **Line 18**: *Request* message for vehicle position.
- **Line 19**: Vehicle position is transmitted.

4. Conclusion

With all previous programming examples installed and explained, you should now have a pretty good grip on programming your own SAE J1939 application with the Arduino Uno and/or the Mega 2560.

Yet, there are still numerous, possible scenarios for J1939 applications and all of them may bare their own challenges. However, this book with its hands-on approach should help you to avoid an otherwise steep learning curve. I believe, that the included screen shots of the SAE J1939 data traffic are a tremendous help to understanding the J1939 protocol.

As I mentioned previously, the main focus of this book was the implementation of a fully functional SAE J1939 protocol stack using the Arduino hardware. I have not covered the standard Arduino input/output capabilities and the numerous sensor-reading applications that come to mind. Such applications will most certainly require additional hardware, specifically Arduino shields, and they will all come with some time-extensive research of their capabilities.

At this time, I will leave the discovery of more sophisticated applications to you but will contemplate writing another book on the topic.

Have fun!

Appendix A – Debugging Macros

Please be aware that the following macros are **one-liners**. Due to the limited page space, some macros appear as wrapped into more than one line and the line breaks are indicated with "\".

```
#define DEBUG_INIT() char sDebug[128];

#define DEBUG_PRINTHEX(T, v) {Serial.print(T); sprintf(sDebug, "%x\n\r", v); \
Serial.print(sDebug);}

#define DEBUG_PRINTDEC(T, v) {Serial.print(T); sprintf(sDebug, "%d\n\r", v); \
Serial.print(sDebug);}

#define DEBUG_PRINTARRAYHEX(T, a, l) {Serial.print(T); if(l == 0) \
{Serial.print("Empty.\n\r"); else {for(int x=0; x<l; x++){sprintf(sDebug, "%x
", a[x]); \
Serial.print(sDebug);} Serial.print("\n\r");}}

#define DEBUG_PRINTARRAYDEC(T, a, l) {Serial.print(T); if(l == 0) \
Serial.print("Empty.\n\r"); else {for(int x=0; x<l; x++){sprintf(sDebug, "%d ",
a[x]); \
Serial.print(sDebug);} Serial.print("\n\r");}}

#define DEBUG_HALT() {while(Serial.available() == 0); Serial.setTimeout(1); \
Serial.readBytes(sDebug, 1);}
```

Appendix B - ARD1939 Protocol Stack Reference

The functions available to the SAE J1939 application layer (i.e your program) are:

Initialization

- *j1939.Init* – Initializes the protocol stack settings
- *j1939.SetPreferredAddress* – Sets the preferred node (source) address
- *j1939.SetAddressRange* – Sets the negotiable address range (optional)
- *j1939.SetNAME* – Sets the ECU's NAME using the individual parameters
- *j1939.SetMessageFilter* – Sets the PGNs to be processed in your application

Read/Write – Check Status

- *j1939.Operate* – Handles the address claim process, reads PGNs from the vehicle network, and delivers the current protocol status (Address Claim in progress, Address Claim successful, Address Claim failed)
- *j1939.Transmit* – Transmits data to the vehicle network and handles the Transport Protocol (TP)

Other Application Functions

- *j1939.Terminate* – Resets the protocol stack settings
- *j1939.GetSourceAddress* – Delivers the negotiated node address
- *j1939.DeleteMessageFilter* – Deletes a message filter

Application Structure

As with every Arduino application, the initialization of data and the J1939 protocol takes place during the *setup()* function, while the actual application resides in *loop()*.

- **setup()** - The *j1939.Init* function is mandatory to operate ARD1939. You also need to set the preferred address and a NAME (for test purposes you can use the project default settings). Setting an address range is optional.

  ARD1939 will not deliver any data (PGNs) unless you set the message filter. ARD1939 supports up to 10 (UNO) or 100 (Mega 2560) message filters.

- **loop()** – This function must start with a *delay()* call, directly followed by *j1939.Operate()*. It is mandatory that the *delay* time and the time passed to *j1939.Operate* are identical, otherwise the protocol timing will be off, which, in turn, will cause error conditions.

The functionality settings can be found inside the *ARD1939.h* file:

```
// Program Version
// -----------------------------------------------
// 0 - ARD1939-Uno
// 1 - ARD1939-Uno/TP
// 2 - ARD1939-Mega
#define ARD1939VERSION                      2


// J1939 Settings
#if ARD1939VERSION == 0
  #define TRANSPORT_PROTOCOL                0
  #define J1939_MSGLEN                      8
  #define MSGFILTERS                        10
#endif

#if ARD1939VERSION == 1
  #define TRANSPORT_PROTOCOL                1
  #define J1939_MSGLEN                      256
  #define MSGFILTERS                        10
#endif

#if ARD1939VERSION == 2
  #define TRANSPORT_PROTOCOL                1
  #define J1939_MSGLEN                      1785
  #define MSGFILTERS                        100
#endif
```

Function Calls Description

j1939.Init

void j1939.Init(int nSystemTime);

Initializes the ARD1939's memory, baud rate settings, etc.

nSystemTime - This is the loop time of your application in milliseconds. This information will provide the ARD1939 protocol stack a time base to manage all timers required for various protocol tasks.

Ideally, the system time should be 1 millisecond for best performance, however, up to 10 milliseconds should be sufficient for regular network traffic. Any higher values can be used but may jeopardize performance.

j1939.SetPreferredAddress

void j1939.SetPreferredAddress(byte nAddr);

Sets the preferred node address. Default setting is 128 (see *ARD1939.h* file). This function call is mandatory for initializing the protocol stack; otherwise the stack will not be able to send messages into the vehicle network. The preferred address is independent of the negotiable address range, i.e. it can be set anywhere within or outside that range.

nAddr – This is the preferred node (source) address. It should be in a range between 128 and 252. Addresses lower than 128 are allowed, but are regulated by the SAE J1939 Standard.

j1939.SetAddressRange

void j1939.SetAddressRange(byte nAddrBottom, byte nAddrTop);

Sets the negotiable address range. The default range (per *ARD1939.h* file) is 129 to 247. This function call is optional, meaning the protocol stack will work only with the preferred address.

nAddrBottom defines the bottom of the negotiable address range.

nAddrTop defines the top of the negotiable address range.

j1939.SetNAME

void j1939.SetNAME(long lIdentityNumber, int nManufacturerCode, byte nFunctionInstance, byte nECUInstance, byte nFunction, byte nVehicleSystem, byte nVehicleSystemInstance, byte nIndustryGroup, byte nArbitraryAddressCapable);

Sets the ECU's NAME by using individual parameters.

The following shows the default settings for the device's NAME as found in the *ARD1939.h* file:

```
#define NAME_IDENTITY_NUMBER               0xFFFFFF
#define NAME_MANUFACTURER_CODE             0xFFF
#define NAME_FUNCTION_INSTANCE             0
#define NAME_ECU_INSTANCE                  0x00
#define NAME_FUNCTION                      0xFF
#define NAME_RESERVED                      0
#define NAME_VEHICLE_SYSTEM                0x7F
#define NAME_VEHICLE_SYSTEM_INSTANCE       0
#define NAME_INDUSTRY_GROUP                0x00
#define NAME_ARBITRARY_ADDRESS_CAPABLE     0x01
```

nArbitraryAddressCapable should be set to zero when your application does not support a negotiable address range (see function *j1939SetAddressRange*).

The NAME fields have been assigned in a way that they will not interfere when used within an existing vehicle network. This has been done by setting the *Identity Number* and *Manufacturer Code* to the maximum value, which will result in a more passive role during the address claim process. An ECU with a NAME of higher value is more likely to lose the competition with another node using the same address.

Note: *All settings as shown are used for demonstration purposes only. In all consequence, you must follow the SAE's recommendations. Also, you alone (and not the author or publisher) are responsible for the final implementation and the results thereof.*

j1939.SetMessageFilter

byte j1939.SetMessageFilter(long lPGN);

Sets the PGNs to be processed in your application. ARD1939 supports up to 10 (UNO) or 100 (Mega 2560) message filters.

lPGN – This is the PGN you allow to be passed to your application.

Function returns OK or ERROR (as defined in ARD1939.h) where EROR means that no more message filters are available.

Special case – Request Message:

As defined in the SAE J1939 Standard, the *Request* message is 0xEA00, where the LSB is used as the destination address, i.e. the address of the ECU that is supposed to provide the requested information (e.g. 0xEA80, where 0x80 is the destination address).

The ARD1939 protocol stack's message filter, however, will allow <u>every</u> message in the 0xEAxx range to pass when you set any filter PGN in the 0xEAxx range.

This behavior is necessary to allow the global address (255), meaning there are scenarios where one ECU requests the information from all nodes in the network.

As a consequence, it is mandatory that the application, in addition to the *Request* message, also verifies the destination address with its own address.

J1939.Operate

byte j1939.Operate(byte* nMsgId, long* lPGN, byte* pMsg, int* nMsgLen, byte* nDestAddr, byte* nSrcAddr, byte* nPriority);

Handles the address claim process, reads PGNs from the vehicle network, and delivers the current protocol status (Address Claim in progress, Address Claim successful, Address Claim failed).

The function returns ADDRESSCLAIM_INPROGRESS, NORMALDATATRAFFIC (Address claim successful), or ADDRESSCLAIM_FAILED.

The parameters passed to the function are pointers to:

nMsgId = J1939_MSG_NONE – No message received or J1939_MSG_MSG_APP – Message was received.

lPGN = PGN of received message

pMsg = Message data array

nMsgLen = Size of message data array

nDestAddr = The message's destination address (usually the source address of your application but could also be the global address 255 – message broadcasting)

nSrcAddr = The source address, i.e. the address of the node who sent the message.

nPriority = Message priority.

j1939.Transmit

byte j1939.Transmit(byte nPriority, long lPGN, byte nSourceAddress, byte nDestAddress, byte* pData, int nDataLen);

Transmits data to the vehicle network and handles the Transport Protocol (TP), meaning it handles data messages between 0 and 1785 bytes long (For the Arduino Uno, this number is limited to 256). The function automatically invokes the Transport Protocol (TP) when the message is longer than 8 bytes.

The parameters passed to the function are:

lPGN = PGN of the message

pMsg = Message data array

nMsgLen = Size of message data array

nDestAddr = The message's destination address (could also be the global address 255 for message broadcasting)

nSrcAddr = The source address, i.e. the address of your ECU.

nPriority = Message priority.

J1939.Terminate

void j1939.Terminate(void);

Resets the protocol stack settings.

j1939.GetSourceAddress

byte j1939.GetSourceAddress(void);

Delivers the negotiated node address; will be NULLADDRESS (254) in case the address claim process failed.

j1939.DeleteMessageFilter

void j1939.DeleteMessageFilter(long lPGN);

Deletes a message filter.

lPGN = PGN to be deleted. Any attempts to delete a PGN that has not been set previously, will be ignored.

Appendix C – Recommended Literature

There is more than plenty and valuable literature available on the Arduino, but, being an experienced programmer, the one and only work I read was:

Programming Arduino
Getting Started with Sketches
By Simon Monk
ISBN 978-0071784221

Also recommended for providing more background information:

Controller Area Network (CAN) Prototyping with Arduino
By Wilfried Voss
ISBN 978-1938581168

A Comprehensible Guide to Controller Area Network
By Wilfried Voss
ISBN 978-0976511601

A Comprehensible Guide to J1939
By Wilfried Voss
ISBN 978-0976511632

All titles by Wilfried Voss can be found at online bookstores such as Amazon and Amazon Kindle (world-wide), Apple Bookstore, Barnes & Noble incl. NOOK, Lulu.com (PDF download), Kobo, Abebooks.com (incl. all international websites), and any other good bookstore.

CPSIA information can be obtained
at www.ICGtesting.com
Printed in the USA
BVOW07s0050070816
458127BV00006B/20/P